GW00722451

Atlantic Gigantic

A BOOK BY COLIN SOUNESS

THE PLACES WE'VE BEEN BOOKS
Chicago

ISBN hardcover: 978-0-9890389-2-8
ISBN paperback: 978-0-9890389-4-2

Library of Congress Control Number: 2015934337

Nature: Ecosystems & Habitats - Oceans & Seas
Travel: Special Interest - Adventure

Note: This book is a work of memoir.
The author has obtained permission from the other two people whom have consented to appear centrally in the book and have provided their log notes.

Published by:
The Places We've Been books/The Places We've Been, LLC
Chicago, IL U.S.A.

THE PLACES WE'VE BEEN BOOKS
Chicago

www.theplaces35.com
www.facebook.com/ThePlaces35

First edition, 2015
10 9 8 7 6 5 4 3 2 1

For my father, Jim Souness

And for the men and women of the Royal National Lifeboat Institution, who put themselves on the line every day, saving lives at sea, all around the UK and the Republic of Ireland

Waypoints

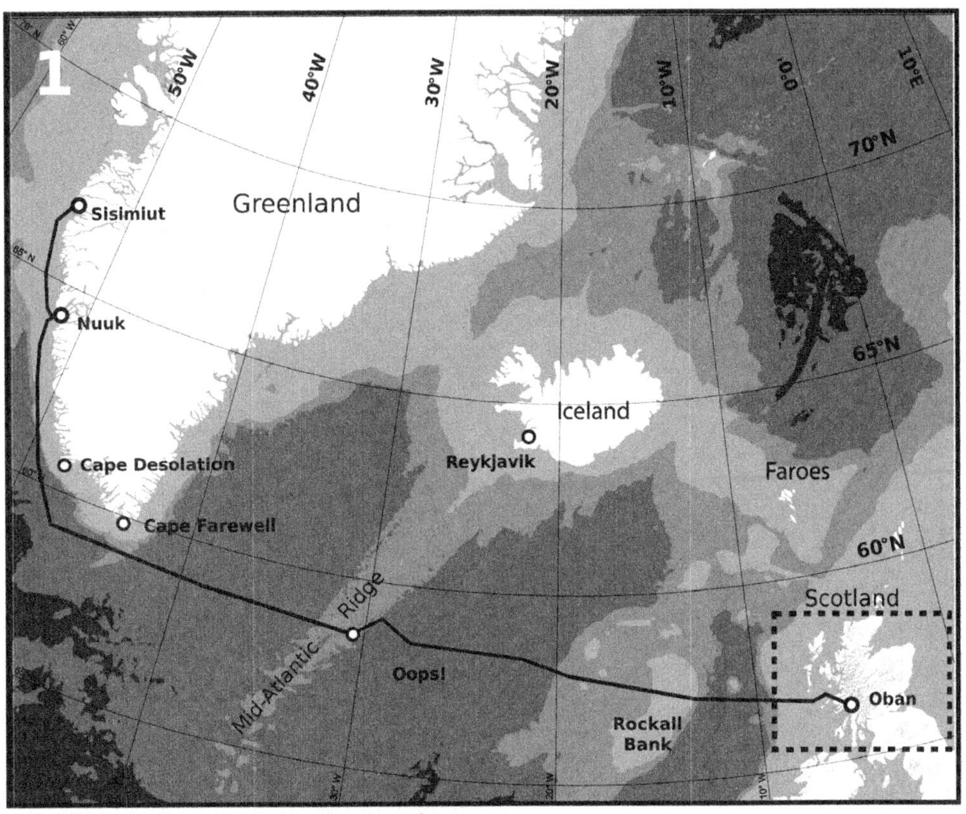

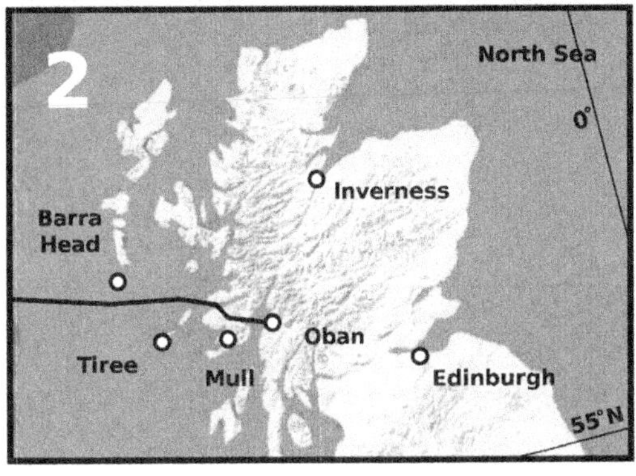

These maps show the approximate route that Gambo took accross the Atlantic on her way from Greenland to Scotland in 2010. '1' shows the north Atlantic area while '2' shows a more detailed map of our eventual approach to Oban in Argyll and Bute, Scotland. The route is shown in both '1' and '2' as an unbroken black line. Several other landmarks and points of interest are shown along the way, including our little "Oops" SSW of Iceland.

Maps adapted from: www.spp-haefen.de

Prologue

Cold and damp I sat in the yacht's cockpit, staring at my booted feet. Beneath them water sloshed backwards and forwards, gently and repeatedly moving the assembled detritus of almost two weeks at sea, back and forward across the deck beneath the duck boards. Pieces of trimmed cord; torn biscuit-wrapper corners; the odd wood shaving; one or two rusted screws; a tea bag; one small metal spoon which no-one had deemed important enough to trouble themselves with retrieving. This varied collection of miscellaneous refuse, now mobilised by the seawater which inevitably collects in the low points of any boat, didn't make for the most exciting of views. But it was somehow relaxing. Somehow, at that moment, the garbage added texture to my otherwise flattened mind. I kept staring downwards. I just didn't want to look up.

A grey, mottled sky stretched outwards and away in every direction, muffling anything the world had to say in a silent explosion of marbled monochrome. Beneath this shroud only the incessant winds moved, lethargically rolling over the mournful deep of the Atlantic Ocean. Waxing and waning, the breeze massaged the surface of the water into a seascape of waves – a world both unchangingly uniform and unfathomably complex. Every wave both seemingly identical and irretrievably unique, as collectively they all tumbled from one fleeting moment into the next. Waves upon waves upon waves, and through them we passed. Again and again those waves broke against grey-painted steel as *Gambo* – our dishevelled-looking yacht – dragged its weather-beaten bulk, slowly slapping through the endless, textured depths. Severed grab lines hung limply from the boat's side. The remains of a shredded sail lay crumpled on its aft deck, forlorn and deflated, like the empty skin of some great white creature; once strong but now slain. The engine – normally the beating heart of a modern vessel – sat cold and motionless. Gambo rolled helplessly, like a scarred, bloodied and exhausted beast, drowning in silence. We sat, three men, crippled in the middle of an ocean. In every direction the shadow-soaked waters of the North Atlantic stretched like a bad dream we couldn't wake up from. Looking out at the horizon seemed to me like peering over the edge of a cliff. I felt dizzy and nauseas and my stomach sat in me like a badly coiled rope, full of knots.

We were over 300 miles from the nearest coastline – Iceland – and things aboard our 15-meter-yacht looked bleak. Most of our largest sail lay in tatters on the deck beside us, mutilated by our own hands. Those parts of the sail which were missing lay a mere five yards or so away from us, below the waterline. Down and out of sight beneath Gambo's hull they were pulled taught, wrapped firmly around the yacht's now motionless propeller. Our mainsail was fully deployed, un-reefed and undamaged, full sheet to the heavens … but in a solidly contrary wind we were languishing at little more than a 2-knot crawl, drifting in what seemed like anything but the right direction.

It was late September. Sam Doyle, Matt Burdekin and I were deep into our ocean-crossing between Greenland's west coast and our destination port of Oban in Scotland. The lion's share of the voyage now lay behind us to the west, but home was still very, very far away. And here, at this point in time, crippled as we were, it seemed further away than ever. How did we even get here? How did we come to be in this situation? It's not what you expect really, when you wake in the morning and first open your eyes, to find yourself stuck on a stricken sail boat hundreds of miles from land in the middle of the Atlantic Ocean. It's more the kind of thing which happens on the television or in movies, and then only to fictional characters. But no, not this time. This time it was three real people who were sat huddled together in the open cockpit of a handicapped yacht, drifting in the middle of one of the world's mightiest oceans. No engine, no foresail, and a colossal blob of canvas making damn sure we didn't move anywhere fast. Options seemed thin on the ground. We were barely moving at all, and when we did make headway it was in the wrong direction. The prevailing winds seemed set on stopping us from making ground towards home. We couldn't even work with the laborious but matter-of-fact option of tacking homeward; cutting back and forward from southeast to northeast, into and against the wind. With no foresail we were restricted to sailing abeam to the breeze (which was a stoic easterly), almost a full 90 degrees off course. This was compounded by the fact that, at only a couple of knots, we didn't have enough power to change tack. Even if the wind was to come about 180 degrees and push us along in the right direction, in our present sail configuration and with the remains of what used to be our biggest sail trailing limply from the prop, it'd take us weeks and weeks to get home. No, it's fair to say that things were looking about as promising as trench warfare.

The blob of canvas had to go. We needed the engine back. We needed to cast off this unwanted sea anchor and free the propeller. Then, perhaps, we could at least motor to the nearest haven in Iceland. But how the hell were we going to un-foul the propeller? It lay a meter-and-a-half below the waterline and a full 2 meters forward of the stern. This wasn't a lazy harbour either. It was the North Atlantic! But, it had to be done. We had to get down under the hull.

So, there we sat, huddled in the cockpit, all staring at the tired, rusty and salt-caked carcass of what claimed to be a scuba compressor. We had an air tank, a breathing regulator, a diving mask, a dry suit, and here, in front of us, the machine which, if we could coax it into life, should allow us to fill the tank with air. The only problem: none of us knew even the first thing about scuba diving.

Long moments passed as we sat, and stared, and sat some more. The only sounds were those of the water lapping incessantly against the hull, and the awkward, seemingly embarrassed flap of the mainsail as it tried in vain to drag us forward. It was Sam who spoke first.

"Are we really going to do this?"

I broke eye contact with the floor and glanced over at him, relieved that someone had finally spoken.

"I don't think we have much of a choice!"

Another few moments passed in silence as our words drifted away on the breeze, seemingly absorbed by the vastness of the sea and sky. Sam glanced up at Matt and I, and then downwards again at the flaky, grease-caked motor of the compressor. He shook his head ever so slightly as he opened his mouth to speak again:

"That's committed, boys."

1 / The call of the waves

I remember the exact moment when I eventually decided that sailing across the Atlantic might be a good idea and not just a strange, unconventionally wet form of self-harm. It was late spring 2010 and at that time I owned a 125cc Honda 'Rebel' motorcycle. I didn't have a car. Just a bike. It seemed like the sensible option at the time despite the limited cargo capacity. I was always amazed though, at just how much you can fit in an 80 litre Bergen and a set of panniers and still ride in a straight line.

The bike was a laid-back, subdued 'chopper'-style number, replete with 'L' plates. I'd bought it for a mere £500 from a guy who lived near Inverness in the north-east of Scotland. It had been a 'pick-up only' sale which, given the seller's location, explained the unusually low price tag, in my mind at least, as the north of Scotland isn't known for its motorcycle-friendly weather.

I'd stripped the fuel tank down and re-painted her with a somewhat optimistic flame graphic, giving the machine the look of a bike five times her grunt and, importantly, hiding the somewhat embarrassing 'Rebel' decal that had previously labelled her sides. Even if the name summed up the situation (and the kind of person likely to buy one of these bikes) quite concisely, there seemed little need to spell it out in three-inch type. Anyway, there I was, aged twenty-seven, living my second year PhD student's 'mid-life crisis'.

It was a sunny but cold morning and I was riding through North Wales to go climbing at Tremadog. The rest of the gang were going by car and camping overnight later, but as I had commitments back home early the next morning I'd opted to take the bike so as to make it a day trip. This was despite the fact that it had been about minus 5 degrees Celsius overnight. As I blitzed along the seriously bike-happy Welsh roads at a fearsome 50 mph the wind chill began to do its thing and, clad as I was in my open-fronted biker dome and discount biking gear, I got a bit chilly. It was as my beard started to freeze to my helmet, preventing me from moving my jaw without causing myself severe pain, that I started to think that maybe, just maybe, I was enough of an idiot that crossing the North Atlantic in a 15-metre, wind-powered metal tube (i.e. a steel-hulled sailing yacht) might be quite good fun.

The opportunity had arisen through a researcher at Aberystwyth University where, since 2009, I had been studying for my PhD in glaciology. The bloke in question was a sailor and mountaineer, turned scientist, who also happened to own a sailing boat and also happened to be managing a long-term research initiative in the west of Greenland. The boat, named 'Gambo', had spent the last few years of its quirky career in South America and the Antarctic but had recently found gainful employment as a research support vessel, striking close in against glacial calving fronts, penetrating deep into Greenland's breath-taking fjord lands and generally facilitating all kinds of the heroic physical geographising that had lured me into academia in the first place. Now, after a couple of years intermittently chilling out in Greenlandic waters, Gambo was to make her first-ever crossing of the Atlantic and come 'home' to the UK. I say 'home', but in fact she had never once soaked her hull in British waters. Her owner, one Alun Hubbard, was Welsh by birth and had owned Gambo for nigh on twenty years, but never once had she tasted Blighty's brine. She had been built in the west of Canada some thirty years previously and still carried the registration mark

'V.C.' which denotes a boat registered in Vancouver, British Columbia. This code, along with her relatively recent battleship-grey paint job, gave her a somewhat formidable and slightly paramilitary look.

Gambo had previously sported a green hull. This colour, topped by white gunnels and combined with the fierce, red Welsh dragons which had always adorned the port and starboard bow sections, had left few passers-by in any doubt about the nationality of Gambo's owner. This look may also have given spectators something of a taste for Alun's attitude towards life, for green is believed by many seafarers to be an unlucky colour for any ship. This always amused me more than anything else, but as with so many things, the more time you spend at sea the more stock you place in old superstitions. So, although grey perhaps isn't the best colour for a boat if you ever want to be seen by other vessels, at least it isn't green! Perhaps this would be a good thing for Gambo? Time would tell.

So, this season, Gambo was to sail for Oban in the west of Scotland. Here she was to spend the northern hemispheric winter of 2010/2011, before helping out on various other projects in the UK and, ultimately, heading back to Greenland. That was the plan anyway. However, in spring 2010, as I was skidding along the un-seasonally icy roads of Ceredigion, west Wales, Gambo was still in Greenland. An ocean lay before her.

Alun needed crew. The crossing was scheduled for late September 2010. He himself was too thinly spread across other responsibilities to undertake managing Gambo in person. So, ever one to share good, old-fashioned, character building opportunities, he had enlisted a young Frenchman by the name of Nolwenn Chauche to handle Gambo for the duration of the Greenland enterprise. This would, of course, include her foray across the North Atlantic and into British waters. I knew Nolwenn pretty well, having sailed on Gambo with him in Nova Scotia, Canada, earlier that same spring just as the boat was preparing for the coming summer's ventures in Greenland. He was an excellent sailor and a solid friend to have alongside in any salty situation. I had a great deal of respect for him, not least of all because he simply exuded authority at sea despite being four years my junior (remember I was twenty-seven at this time) and also because he suffered from what would, in my case, have been cripplingly bad seasickness! Suffice to say that I liked sailing with him, and that the prospect of crossing one of the world's stormiest oceans seemed considerably less daunting knowing that he would be in charge. Nolwenn also, despite being a consummate professional and a responsible skipper, had a very open-minded attitude to risk and dangerous undertakings. As an open-ocean sailor you really need to.

So, anyway. Back in Wales.... It's spring 2010 and I am careering through the chilly morning air on my somewhat under-powered and over-burdened motorcycle. As the ice set in about my face, somewhere on the tree-flanked roads near the west Wales town of Machynlleth, I decided that sailing from Greenland to Scotland might not be that far outwith my sphere of capabilities after all. This feeling never went away and about six months later I was en route to Greenland to meet up with Gambo and, I thought, Nolwenn and the rest of his crew. By this point it was mid-September. The nights were beginning to draw in and summer was becoming little more than a memory.

This was all a bit of a leap for me, even in principle. I knew from previous experiences on Gambo how uncomfortable prolonged ocean crossings can be in small yachts. I also knew, what with it getting increasingly late in the season now, that this particular crossing was likely to be a difficult one. Still, that's precisely the reason I wanted to do it. I wanted to know if I was equal to the challenge of the North Atlantic. I'd grown up in Argyll and Bute; a region in the western highlands of Scotland and a neighbour to the vastness of the open ocean. I had also spent a lot of time in the islands of the Outer Hebrides over my life. In those flung-out and beautifully exposed places I often found myself working on a sand dune or just standing on a headland looking out westwards at the ocean rolling in. At those moments it always struck me in the gut that out there, beyond the waves, were yet more and more waves. Beyond those, at some unfathomable distance: America. The scale seemed monumental. The Atlantic boggled my perceptions of distance and exposure. With modern air travel it seemed to me that so many people nowadays don't give these distances any thought at all. Whether or not that is true I just don't know, but that's how it played in my mind's eye. Crossing these expanses – taking the time and the risks required to do it independently and on the water – seemed to have become a challenge against which we no longer measure ourselves. Distance is rarely ever a barrier now where money is forthcoming, so people measure themselves against the cost and the wealth required, but not against the physicality of it all. I could never help it though: I always wondered if I was equal to those who had gone before and who had pitched themselves against nature's great barriers. Had my being born into *this* generation and not into one of times past; my social padding; our handy health service and late Twentieth Century upbringing.... Had all these factors made me less robust a being than my ancestors of centuries gone? I hoped and thought not, but could never really know until I took the plunge.

Well, here was my chance....

2 / Like an iceberg in your life

I was now standing upon the very edge of fate's diving board. My stomach was in my throat and my money where my mouth was. I'd packed the bits and pieces I thought might help me through (a good multi-tool, several pairs of Royal Air Force issued trousers, a Scotland football strip, a Hebridean and Clyde Ferries sweatshirt, a waterproof bivvi bag, and a generous measure of open-mindedness) and set off for Gatwick Airport. That's where things took the first of several unexpected turns.

In August 2010 a very large iceberg, about 260 square kilometres in area, broke free from Petermann Glacier. Petermann is one of Greenland's largest and most northerly glaciers. It is a tidewater glacier, meaning that a large portion of its length floats in the deep waters of north-west Greenland's fjord systems. That summer, a great deal of attention was focussed on Petermann. Greenland, and the outlet glaciers of that country's ice sheet, particularly those in west Greenland, were all under observation as the eyes of international science strained to better understand the spectre of climate change, nervously searching for signs that perhaps the world might be changing under our very feet. Teams of glaciologists and climatologists flocked northwards looking for opportunities and case studies that might bring us closer to understanding that change.

Petermann Glacier had been under scrutiny. Like the Martians in H. G. Wells' *War of the Worlds,* Earth's scientists had been watching from afar; looking; wondering; waiting. Sea temperatures had been steadily rising in the Arctic over the last two decades and several other smaller tidewater glaciers in the area had already responded by retreating markedly, dropping, or 'calving', large icebergs into Nares Strait, north of Baffin Bay. Petermann Glacier had a large floating, or 'tidewater' snout, and as average sea temperatures around tidewater glaciers increase beyond 0 Celsius, ice masses such as Petermann tend to become more and more prone to retreat. Basically, there were a few scientists who were pretty certain at the outset of the 2010 fieldwork season that Petermann was due for a major calving event. As chance would have it, a matter of weeks before I was scheduled to fly out to join Gambo in Greenland, Petermann provided the world with the spectacle everyone had been waiting for. The berg which popped off and subsequently bobbed its way out into Nares Strait was equivalent in size to three Manhattan Islands joined side-by-side.

When this massive calving occurred, almost on cue in fact, there was a bit of a dash to get there and take full scientific advantage of the event, both to better understand the processes responsible and to increase the media profile of the work being undertaken on the issue of climate change. Of course, scientific endeavour was the primary reason for Gambo being in Greenland at all, and thus the reason why I was heading out there now. Therefore, when Petermann kicked up a fuss at her warming waters and decided to throw her icy toys out of her Arctic pram, Gambo upped anchor and scurried off northwards to have a look.

In fact, Gambo's main job on this northerly jaunt was to drop off a substantial amount of aviation fuel which would make it possible for helicopters to get to Petermann, operate in the area for a time, and still have enough fuel to get south again afterwards. It was a supply run, but an important one.

Now, Petermann Glacier is rather far north; 81.1 degrees north to be exact. Up until that point in time Gambo had, for the most part, been in the vicinity of a town called Uummannaq at around 70.1 degrees north. Sailing between these two places is a matter of some 800+ nautical miles! That's over 1600 nautical miles as a round trip, which at a motoring speed of between 5 and 6 knots is easily a few weeks of Gambo's time. So, by early September it looked like Gambo's move into the North Atlantic and back to Scotland was going to happen a little later than had initially been planned.

Of course, a late departure would have some implications for those who were due to crew Gambo. The later in the season we were able to leave it before throwing ourselves on the mercy of the Atlantic's autumnal weather systems, the more of the ocean's fine meteorological cuisine we would be able to sample. Any seafarer worth his salt knows that the most exciting weather is to be found either side of winter.

I'll be honest, as I sat at Gatwick, mid-September, writing my farewell text messages, I hadn't yet fully grasped the significance of this delay in our sailing, neither in terms of weather nor of crewing.

Unfortunately Gambo's protracted commitments north of the Arctic Circle meant that her skipper, Nolwenn, who was about to begin a Master's degree back in France, would miss the start of his course. Understandably he wasn't keen to do this.

So, I am standing in London's Gatwick Airport. I have checked my bag, passed through security, sent my cheery byes to my nearest and dearest (just in case) and am simply biding my time before boarding the plane….

My mobile phone rings. Who should be calling but Alun, on a satellite phone from somewhere in Greenland. He has no crew left. The skipper needs to go home to university and the first mate has badly injured his hand (kneading bread dough, of all things!) so is also done for the season.

"So, I need you to be skipper," said Alun.

"I think I can guarantee you at least a couple of crew. I'm not sure who just yet, but I'll find some. Are you up for it?"

Unbelievable.

I started sweating. I think I probably turned a little pale. I was standing against the railings of the upper tier of the shopping esplanade in Gatwick's duty-free area. All the world and their collective entourages were milling around and beneath me heading god knows where, and I had just been asked if I was confident enough not only to crew a trans-Atlantic voyage, but to actually skipper it. The answer in my head was a resounding no. *No! Definitely not!* I was suddenly convinced that I barely knew how to sail. I had no idea how the boat worked, where I was going or even why she floated. If I was to say yes now I'd never see home again, surely.

So, naturally, it being one of those crossroad moments of 'yay' or 'nay', right now, I said the only thing I could say and subsequently ever be happy to look in a mirror again: "Yes", of course.

Oh for f**k's sake!

3 / Stepping out 'into the wild'

I had the rest of my flight to Copenhagen to dwell on what I had just committed myself to, and then an overnight stop in Copenhagen Airport to revise those thoughts.

Incidentally, Copenhagen is probably my least favourite airport to sleep in on account of the hard metal arm rests which prevent you stretching out along seats, and also the formidable level of cleanliness which demands that floor polishers work loudly throughout the night. These days, when travelling, I always take ear plugs, and when sleeping in airports I generally come armed with a bottle of wine. The combination of the two gives you a good fighting chance at a solid night's sleep and also makes people-watching a whole lot more entertaining.

With the dawn came my flight to Greenland, and so I stepped somewhat tentatively from European soil onto the Air Greenland boarding gangway, consumed with a combination of apprehension and excitement. Now, taking my seat on that big red Airbus and knowing what I would have to do in order to get home again … somehow I couldn't bring myself to assume that it would all go to plan. I was nervous. *Very* nervous. Scratch that, I was terrified!

My fears were an interesting parody of the feelings I'd had when I *first* visited Greenland, back in August of 2003. Travelling to Kalaallit Nunaat (the name by which Greenlanders call their own country, meaning 'the land of the people') for the first time, aged twenty-one, had been not only my inaugural visit to that great frozen island, but also my first visit to the Arctic, and I had felt like I was venturing into a mystical land. Crossing the Arctic Circle was to me like overcoming some existential barrier. Even back then I had travelled quite a lot in my life, both alone and with others. But, Greenland was, at that stage, my most adventurous undertaking so far. My imagination was still feeding on the yolk of high school geography classes; videos and grainy 1980s photographs of perpetually frozen lands where all food must either be shot, netted or somehow harpooned. I had no real understanding of where I was going, travelling on a shoestring budget to the Arctic.

As a twenty-one-year-old undergraduate, still eking by on the remainder of summer earnings, I'd naturally had no money for accommodation so would be camping. And, to top it all, I was going it alone. Then, in the penultimate year of my studies at Edinburgh University, I had seized the opportunity to combine my work with travel. It was time to begin my dissertation and, being well on my way to becoming a glaciologist-in-waiting, Greenland had almost jumped off the map at me. I had begun planning some seven months in advance and had finalised everything only a month later. Thus, I had spent the last six months prior to leaving simply waiting and growing increasingly nervous. Finally, August had arrived and I had set off from Edinburgh, flying cheaply via London to Copenhagen; Air Greenland's only European terminal. As my ticket from Denmark to Greenland was neither cheap (even with a student discount) nor flexible, I was utterly terrified of delays that might result in me missing my flight to the Arctic. This over-anxiousness was the cause of two less-than-pleasant, one-man airport slumber parties. All those months earlier, back in Edinburgh, it had seemed a good idea to arrive almost a day early for every connecting flight, just in case. I had not even reached Greenland yet and was already exhausted!

Needless to say, after two budget flights and two equally budget nights (recall my earlier comments about Copenhagen Airport), embedding myself in my chair aboard Air Greenland's vivid scarlet liner had been as satisfying as a hot bath on a cold day. By that stage in the game, rather than watching the clock with nervous dread I was positively excited about the prospect of a hold-up. Unfortunately the flight takes a mere four hours and I knew that in a short while (far less than the time needed for a solid nap) I would be stepping down from the plane and out into the rugged hills of West Greenland. I would breathe Arctic air for the first time.

For the next twenty-one days my home was to be a small yellow tent with space only for me and my bag. I really had no idea of what to expect, so rather than nap I chose to remain awake and enjoy the comfortable chair and the in-flight television while I could. In any event, sleep would likely have been impossible as by this time I was saturated with excitement.

In what seemed like no time at all the dark blue patches of Atlantic loneliness that had previously been visible through the delicate, tattered blanket of the clouds started giving way to unbroken white. We had reached the vast dome of the Greenland ice sheet. More than 1.7 million square kilometres of what looked like icy nothingness: a solid skin of hard-packed snow layers, more than a hundred-thousand years in the making.

Now, fast-forward. In 2010, just as back in 2003, as I headed once again to Greenland (this time to join Gambo), the clouds dissipated and I was flying across what looked to me like a white ocean, although it was in fact the Greenland ice sheet again; a funny perspective given how I was set to return home. Out over white, home over blue….

In every visible direction the ice stretched away to a lightly-blended, visibly-curved horizon. The view felt deeply personal. This was where I wanted to be, and now I was aching to reach out and touch it. Frustration bubbled in me like a poured soft drink. I knew the ice sheet was huge, but from up here, rather than better appreciating its size, I felt disassociated from it. Even from a jet, travelling at inhuman speeds, it was almost impossible to discern the passing of the surface below. There is no scale when viewing landscapes like this; no features upon which to fix your gaze. It's almost hypnotic.

Studying glacial landscapes and reading old expedition accounts has, throughout my life, made me somewhat reverential of these distant high-latitudes where ice dominates and glaciers abound. As a young man I had grappled with the science of glaciations and the incredible, incremental way in which these relentless rivers of ice carved the Scottish landscapes I had been brought up to love. Through my eyes – the eyes of an exploration-hungry schoolboy – places like Greenland had held a kind of inaccessible celebrity and mysticism. The first time I arrived there, as the ice stretched out below me and the plane began to descend, I felt a very real sense of stepping into a fantasy. Somehow though, as I now returned to Greenland as a grown man, the business end of the feelings had changed shape. I felt less like I was here to explore and more like I was here to fight.

4 / Grounded helicopters, flat tyres and Russian billionaires

On arriving at Kangerlussuaq (the town which houses Greenland's main international airport, and very little else besides) the plan was for me to track down a certain Irish helicopter pilot. This pilot had been flying helicopter support for research projects which, up until only recently (this was now the end of the summer fieldwork season), had been ticking over, up on the ice sheet. The pilot's name was Martin and it was he who would be flying me north to join Gambo, all going to plan that is.... I was under the impression that Martin was expecting me and would meet me at the airport.

Now, all my life I have wondered what it would be like to arrive somewhere by air and be met by a stranger holding a piece of paper with my name on it. A few times I've expected it, but on every occasion I've ended up wrangling public transport instead, which is its own kind of fun. As I arrived into Kangerlussuaq I had a feeling that my track record on these matters was too good to be broken. As things happened I wasn't to be disappointed. No-one had even told Martin that I was on my way!

Eventually, and perhaps inevitably, I found myself alone in the Air Greenland arrivals room. At this stage I didn't really know how long I would be staying in Kangerlussuaq, but regardless of the time scale I was fairly sure I'd be happier somewhere other than the airport lounge. So, as it's a small town, I somewhat nervously started asking around the various information desks and tourist stalls whether anyone knew, or had seen, an Irish helicopter pilot recently. Virtually everyone seemed to know Martin (I think Irish folks are something of a minority in Greenland), so I managed to figure out where he was staying. This was a start. Unfortunately, however, the one or two people who'd known his phone number and called him for me had only managed to get an answering machine. So, as I'd spent more than enough time in planes and airports already on this adventure, I decided my best course of action at this point was just to start walking.

Martin was billeted in 'Old Camp', which is a collection of prefab wooden huts located some 2 miles to the west of Kangerlussuaq and used as a kind of transitory village by visiting aircrew, scientists and tourist groups. After about half an hour of walking down memory lane (it'd been out this way to the west of the airport that I'd camped when staying in Kangerlussuaq seven years previously) I arrived amidst the sun-bleached, red shacks of Old Camp. I found the one which had been let out to 'the project' and gingerly crept inside looking for signs of life. There were a couple of 4x4 trucks parked outside, so I figured I had a good chance of finding my man.

Inside the hut it was silent, but from the smell of stale beer and men's deodorant I felt sure I was in the right place. Here be pilots!

I eventually found Martin sound asleep in his bunk doing his best to overcome the vestiges of the previous night's hangover. He was none too pleased to meet me it has to be said, but in fairness it was early on a Saturday morning….

After a cup of coffee or two (we both needed one) Martin pointed me to what he assumed must be my bunk (it was the only empty one) and I set about establishing myself for what I expected and hoped would be a short stay. As the lingering clouds of Friday night lifted from the cabin, a far more jovial atmosphere began to settle in and as Martin began to emerge from his Saturday pre-noon shell he started moving even closer to the profile I had already preconceived about what an Irish helicopter pilot in Greenland might be like. A bit of a legend really!

There were two pilots and two helicopters working on Alun's project and it was his intention that one of these choppers would be heading north, up towards the recently calved Petermann iceberg. I was to ride shotgun so as to join Gambo's motley crew and get stuck-in about boat-y stuff early, before the skipper and everyone with any sailing competence (i.e. everyone other than myself) left. The chopper would rendezvous with the boat and drop me off before she began the southbound return leg of her operation. This all seemed typically Gambo: Most vessels these days are pushing into the high Arctic in *search* of petroleum resources. Always one to fly against convention, Gambo's objective was, naturally, to import and *store* petroleum resources. Hiding petroleum in the high Arctic is, of course, considerably more straight-forward than *finding* petroleum in the high Arctic. However, as Martin soon related to me, there was a problem. A couple of problems actually; both of them with the helicopters.

One of the helis had a mechanical issue. She had therefore been grounded until repairs could be carried out. The other helicopter which was flying research support was in good working order but, unfortunately, the pilot whom was licensed to fly that particular airframe was unable to secure approval for the northbound flight plan which had been submitted to the Greenlandic government.

So, we had two aircrew and two helicopters. The pilots were non-interchangeable between airframes due to the rules surrounding operating permits. Therefore, while we had a working chopper *and* a permitted pilot, the two couldn't be brought together legally, and thus we had nothing. From where I stood, this state of affairs put me in the typically Arctic position of having nothing in particular to do, but also having no freedom to stray beyond the 'city limits' in case by some contrivance of chance the 'broken-down' helicopter was to suddenly fix itself. I was on standby in a town with only two shops, no real entertainment facilities, where beer cost a prohibitive seven pounds per bottle, and the highlight of every working day was the arrival of the incoming flight from Copenhagen.

If there was ever a place on Earth where you could go to build upon your eccentricities, Kangerlussuaq is that place.

In places like 'Kanger'; tucked away at the head of an Arctic fjord; a stone's throw from the Greenland ice sheet; where everyone passes through but no-one stays; where musk oxen are frequently seen roaming the streets between the various hangers, sheds and industrial lean-tos which skirt the runway; where the social hub of the town is the coffee table in the regional flight control office.... In places like this pretty much everything seems touched by the brush of the surreal. It was all a little *'Twin Peaks'* really, and I had a few escapades in Kanger which I think are fun tales in their own right. You'll find a couple of these mini-adventures (i.e. the 'flat tyres and Russian Billionaires' component of this chapter's title) snuck in as stand-alone 'bonus features' at the end of this book.

5 / The crew musters

As the days turned into weeks I grew increasingly comfortable in the sedate, aero-centric little world of Kangerlussuaq. However all things pass and eventually Alun conceded, via email, that it seemed unlikely any choppers were going to fly northwards anytime soon. Gambo was now well on her way back south again anyway, so it was decided that I'd be meeting her in the relatively nearby fishing town of Sisimiut (about 100 miles distant to the west). A date was set and a flight was booked for me on one of the small Dash turboprop passenger planes which provide most of the inter-settlement links in Greenland. It seemed my days in Kanger were numbered! This was a shame on a couple of notes: Firstly, I was now getting to know people. I'd even gotten myself a job. An unpaid job it must be said, but a job nonetheless. I would be painting a wooden holiday home recently built by the Danish manager of the town's dominant (to the point of being virtually nationalised) maintenance firm. I simply went into their hanger one day, pleaded boredom verging on sloth, begged for some kind of distraction and was handed a set of overalls, several vats of wood treatment, several more of yellow paint, a warm jacket and the keys to a

Jeep Cherokee. Happy days! I managed to kill the best part of a week on that little project and certainly enjoyed the freedom of having a 4x4 at my disposal for the duration. I almost began to feel like a local!

The second downside of learning that I'd soon be leaving Kangerlussuaq was that it brought into sharp focus the inescapable reality of why I was here in the first place, namely to skipper a small(ish) sailing boat across the North Atlantic. In the autumn. I once again began to get 'the fear'.

As the landscape around Kanger started to undergo the alarmingly rapid transition from summer greens through intermediate yellows and finally into striking autumnal reds, my thoughts turned to the weather, the waves, the winds, and the ever pervasive salty wetness that define trans-ocean sailing. I'd been in Kanger for over two weeks. Where would I be in two more?

My contact with Alun had been sporadic over the previous fourteen days but I'd learned that he'd managed to secure at least one other crewman for Gambo. This was a relief. The most recent recruit's name was Sam Doyle; a fellow PhD student, originally from Sheffield, England, but who was also studying at Aberystwyth.

It's a fact that, at least at our university, projects in glaciology seem to attract a certain type of person. That person, if you'll forgive me for generalising, is physically outgoing, wilderness minded, has a passion for travel (sometimes verging on an obsession) and rarely manages to say no to a good challenge. In my mind at least, Sam was the veritable exemplar of this stereotype. From his general competence and easy-going confidence in the field you would be forgiven for thinking that Sam was a fully-fledged member of the university's staff (Sam had essentially been running the field camp here in Greenland for most of the season thus far), but in fact he was only just beginning the first year of his project and was several years younger than me. I had of course met Sam before, but what with both of our peregrine ways, my time spent overseas earlier in the year (sailing Gambo in Canada) and the fact that Sam had spent most of the summer thus far in Greenland, we didn't really know each other all that well. We had a lot in common, what with both being keen climbers, both being glaciologists and both having a very physically-oriented, outdoors attitude.

But, despite all the groundwork being there we hadn't really had the chance to become mates yet.

Now, over the course of the years following on from this adventure, Sam and I would spend quite a lot more time together. We were flatmates at one point, climbing partners on numerous occasions and even frequent caving buddies. In all that time we became closer and closer as friends, and I think we understand each other pretty well. We share a number of values and, I believe, more than a few underlying motivations in life. But, if there's one aspect of Sam's character to which I cannot relate it is his persistent penchant for understatement. It is truly phenomenal. Sam is the kind of guy who could sit on a beach watching the most outstanding sunset, being fanned and fed cocktails by stunning supermodels while a hundred dolphins breached and cavorted in the sun's final rays and then turn to face you with only the slightest hint of a smile on his face and say something like:

"Quite a nice night, isn't it?"

Aboard Gambo, we all kept a record of events in the ship's log but Sam was by far and away the most assiduous in this—and so I quote him often. (In his aversion to literary exuberance, when you can read between Sam's lines, it's hilarious.) Although I don't always relate to the thought processes behind this subtlety of Sam's, I certainly enjoy it.

I think perhaps one of the reasons Sam fosters this attitude, and I might be wrong, is because he enjoys surprising people. Sam once shared a great example of this predilection for the subdued with me, about how midway through our Atlantic crossing he had sent a happy birthday text message to his father, and that in that text message he had said that he would be a bit late in returning to the UK as he had decided to sail home rather than fly. That is all he said. And so, lacking any further context or any hint of the emotion one might expect when a family member sends you word that they have embarked on a potentially life-threatening adventure, Sam's father quite understandably assumed his son was catching a ferry or some other passenger vessel home. Sometime later, after we had (eventually) made it back to the UK, Sam's dad, finally privy to how we had actually travelled, exclaimed to Sam in some frustration: *"It's not Rother Valley Sam! It's the bleedin' Atlantic!"*

Rother Valley is a sedate, family water-sports park in the north of England.

Now, during the writing of this book I asked Sam to send me through a short biography, so as to properly introduce him. He wrote me a great account of what he considers to be his first ever adventure, giving an insight into how his gravitation towards unconventional exploits began. I'd like to share that here, and let Sam introduce himself through it. I think it paints a nice (and typically understated) picture of a boy growing up without fear of stepping off the edge of the map:

My earliest memory of an adventure is a cycle ride to a park in Sheffield. I was with my friend Gavin. We were both aged only twelve and this was a new park to us, some twenty minutes from our homes. We explored the park and decided to carry on, cycling further and further, eventually into the Peak District National Park. I would cycle this same distance now without batting an eyelid, but back then it was a different story. I was pushing my limits, expanding my boundaries and meanwhile getting very late for my dinner. It was great.

After a long ride we descended down the other side of the hills and into a town called Grindleford, where my Auntie Alison lived. We were exhausted, but felt jubilant, albeit with an edge of trepidation, for my auntie was more than a little surprised to see us. She made me call home and tell my mum where I was. Unfortunately my mum didn't believe me. Alison had to take the phone and assure her that I was telling the truth and really was where I said I was. After that Alison gave us a lift back to the top of the hills in the direction of Sheffield and from there we freewheeled all the way home.

Many biking trips followed this one, and the adventures continued. Next I learned how to kayak (and then subsequently how to swim).

So, Sam was never shy of taking on a challenge, with or without prior planning. Fast forwarding to Greenland in 2010, the story of how Sam found himself 'upgrading' from his return plane ticket to a place on the 15-metre, steel salt trap that I'd ostensibly assumed responsibility for is a fun one and echoes his memories of childhood misadventure nicely. It also details how the third member of our little pelagic posse came to sign up.

As I've mentioned, Sam had been working in Greenland under Alun's direction for most of the summer thus far. Being an opportunistic and adventurous chap Sam had made the most of this arrangement and, arranging in advance with Alun that he would take a few weeks off towards the end of the summer fieldwork season, had managed to organise a climbing expedition for himself and some of his friends from back home. Their aim was to put some new 'trad' rock routes up the impressive sea cliffs and mountains around Uummannaq – a smallish Greenlandic town which lay close to Gambo's main area of research activity. He'd even managed to secure some pretty respectable funding which, combined with the use of Gambo as a support vessel (i.e. climbing taxi and emergency getaway vehicle) made everything possible. Sam's friends Miles Hill, George Ullrich and Matthew Burdekin flew out to Greenland and, together, they achieved some pretty cool things.

By early September, with their expedition coming to an end, these lads were set to go their separate ways. Sam was to remain for a short while and assist in the closure of the university's field season while George, Miles and Matt would leave Uummannaq by helicopter, transfer to Ilulissat and then fly home to the UK via Kangerlussuaq and then Copenhagen.

Now, as we know, it transpired that a particularly large iceberg lay in the path of existing plans. As the Petermann berg calved and Gambo rushed off north to get in on the action, it quickly became apparent that everything Gambo-related would be set back considerably. There would be no skipper and there would be no first mate. So, crew were needed! Never one to let an opportunity like this pass, Sam stepped up, all this in spite of having virtually no sailing experience whatsoever. But who needs experience? Some might say that adventure can be measured according to its inverse relationship to preparation, and by my calculator they'd be right. In fact Roald Amundsen, the Norwegian explorer who famously beat Britain's Captain Scott to the South Pole, once said: *"Adventure is just bad planning."*

Now, it is also important to note that Sam had no homeward flights booked. These flights are NOT cheap. Aviation fuel is expensive. Wind, on the other hand, is free! So, additionally, when Sam saw the chance for a spot of jolly good, old-fashioned 'against all odds' exploration, blended deliciously with a chance to save hundreds of pounds, he understandably, albeit somewhat understatedly, jumped at the chance to get on board, so to speak. But what of crewman number three you ask? Well, you'll remember that Sam's climbing buddies – George, Miles and Matt – were now preparing to return home to the UK. George and Miles were ready to leave Greenland, but Matt Burdekin (another Sheffield man)... Matt was different.

Matt describes himself as an *"explorer, mountaineer and miscreant"*. He is one of the most energetic and adventurous people I have ever met. Nothing is ever boring with Matt. Everything is unorthodox and seemingly experimental. This isn't due to any lack of planning. Quite the contrary! The adventure of being around Matt comes simply from the fact that Matt is one of those inspiring kind of folks who need the excitement of the unknown to keep themselves interested in moving forwards. Matt told me once about how, as a small boy, he would go walking in the mountains with his parents and brother. If left unattended he would walk a short distance only before stopping and sitting down on a rock. Bored with the way things were panning out he would just sit there and stubbornly refuse to move, hoping that one of his parents would come back and carry him. That would make things a bit more exciting! Obviously though, things didn't always go according to little Matt's plan and he would be left behind until, feeling abandoned, he would eventually jump up and run after the rest of his family who were by this point well ahead. This kind of behaviour remained a feature of Burdekin family outings until Matt's dad hit on a new strategy to keep his son inspired. He would plan walks in advance, making sure to include an interesting adventure, or some kind of mystery, into the itinerary. One of Matt's earliest memories of these 'enhanced' walks is of being made to crawl through a short drainage tunnel – one of the sort which allow rivers to pass underneath roads – in Burbage Valley in England's Peak District. To a little boy it felt like an unexplored cavern. Would he make it through? What if the river suddenly flooded? Would little Matt be swept away by the icy waters? It was things like this, little adventures, which eventually got Matt inspired by the outdoors. Every walk became a mini expedition complete with stories of danger and daring with which young Matt could enthral his classmates back at school.

As an older boy Matt joined the scouts where he met likeminded people with whom to create new and exciting situations of adversity. As he grew older he did more and more sports. Some agreed with him, some didn't, and amusingly, given the watery context of this book, Matt once told me of how he decided that canoeing was a silly idea after having ended up trapped in rapids, drowning, waiting for someone to throw him a rope. But it's moments like that one which for Matt, and I believe also for Sam and I, make life worthwhile. Matt is someone who becomes easily captivated by the emotions of a moment, and pursuits such as sailing, white water kayaking and rock climbing.... These are undertakings FULL of moments; challenging moments; moments of fear; moments of exhilaration; moments of intense challenge and moments of immense satisfaction. The feelings which course through you in these moments can be so intense and so three-dimensional that everything else seems mundane by comparison. They can be addictive. After discovering climbing, school days for Matt became all about when he could next get out and get back onto the rock.

So, in the light of all this, it maybe won't come as much of a surprise that Matt's decision to get involved in Gambo's sail to the UK was made somewhat more dramatically than Sam's. By autumn 2010 Matt, like everyone else, had commitments back home so he genuinely had to think quite hard about his options. Similarly to Sam, Matt is not a bloke who shies away from an opportunity to give the enemy a good what for and make it home in time for tea and medals, a fact which should be manifestly obvious given the reason that all four of these lads were in Greenland in the first place! So, needless to say, when the rallying call went out for crew the thirst for action cut him deeply. But what of his Master's degree? Here he was in Greenland, physically and psychologically very far removed from such seemingly trivial matters as bank statements, tenancy agreements and matriculation dates. This is, after all, part of the joy of exploring remote places: more worldly concerns can be vanished from your mind, at least temporarily. But, he was en route home now, and these matters, like weeds, had begun to take root once again. So, I imagine it was very much in two minds and torn between two worlds of priority that Matt put his packed bags onto the helicopter which would carry George, Miles and himself back to Ilulissat. But even then, as the rotors began to spool up, I think

Matt knew that this particular mosquito of adventure which had settled on him had already bitten and wasn't going to be swatted without putting up a fight. Sam's short but sharp words of encouragement rang in his head.

"Matt. We could sail it back!"

This conversation had occurred as Sam walked his mates to the helipad in Uummannaq. Matt's bags were already loaded. The chopper was a large, red, passenger Sikorski. It sat awaiting its human cargo, rotors already spinning at idling speed. This was it. Crunch time. So – and I really wish I'd been there to see this – with literally moments to spare Matt made his call. With blades flitting powerfully overhead, the chopper's door opened to admit him. But, in a sudden blow-out of spontaneity, Matt opted not to climb aboard.

"Nope, I'm not getting on. I'm sailing home."

The bags were jettisoned back onto the helipad by bemused staff. The decision had been made. Sailing seemed like an infinitely less comfortable way of travelling from Greenland to the UK; how could Matt refuse?

Matt Burdekin quite literally made this decision in five minutes flat. He saved 150 quid by bailing on the flight from Uummannaq. He lost about 800 by skipping the flight home to the UK. But just think of all he stood to gain in the clutches of the Atlantic! Hmmmm….

So, it seemed that at least I wouldn't be alone in being alone on the sea, so to speak. This was good news! However, it did mean that this really was going to happen. Now that we had a practical number of people with which to crew the boat the whole operation was virtually confirmed. In the back of my mind, what with the pervasive melancholy of Kangerlussuaq assaulting my senses for the last two weeks and the almost comical failure of Greenland's aviation infrastructure to get me north, it had seemed more than a remote possibility that I might end up having come all this way for nothing.

I had begun to imagine that perhaps Gambo might just give up on the whole enterprise and resign herself to being frozen into the sea ice for the winter. Now, however, we were on! I was simultaneously relieved, excited and dismayed at the prospect. It was undoubtedly going to be one of the most remarkable experiences of my life, negotiating everything the North

Atlantic had to throw at us without guidance, experienced leadership or good old-fashioned top cover. We would be the only ones to decide what to do, when to do it and how it should be done. I had only ever been basic crew on Gambo up until that time (and I'd only spent two weeks sailing in the last five years), Sam had never sailed a yacht in his life and Matt, who professed to be very prone to seasickness, had only ever racked up substantial time sailing small dinghies. What the hell were we thinking.... I honestly didn't know if we were up to this challenge. Were we tough, adaptable creatures like the men of bygone ages? I didn't know. What I did know though was that we were most definitely going to find out.

As you can imagine, the last thing on Earth I needed at this juncture was lots of time alone. Time spent in my own company seemed likely to result in the full over-inflation of the looming sense of dread that was beginning to blossom in my mind. So, me being naturally myself, what with no longer being on 'standby' and what with my flight to Sisimiut (where I would meet Gambo) being in a full three nights time, I decided that the most sensible thing to do would be to throw the bare minimum of provisions into a rucksack and just wander off into the tundra. Therefore, with four days left before my eventual deportation, I set out with my boots, a book, stale bread, a block of cheese and a bivvi bag in search of a bit of 'me time' in the wilderness.

The nights were truly dark now, and the temperature was dropping off sharply. The very day my movement restrictions were lifted I saw pancake-ice forming, for the first time that season, on the brackish waters of the fjord head. Winter was baying in the distance, and the wind from the inland ice carried her call to my nervous ears.

South of Kangerlussuaq there's a large lake: Lake Ferguson. It was to this lake that I headed initially. From here I would move up onto the raised tundra to the south-east, further inland past numerous other smaller lakes, passing by the remains of several ancient Norse and Inuit settlements and eastwards towards the edge of the vast white ice dome of the Greenland ice sheet which, even if sometimes you cannot see it directly, dominates the landscape in these parts.

As I walked along the eastern shore of Lake Ferguson the daylight gradually began to fade. Evening was setting in, and although the afternoon's sun had been strong and warming, as soon as the already low Arctic light began its inevitable descent towards the western horizon the air temperature started to dip sharply. I needed to settle on a site for my 'scratcher' (British, armed forces slang for one's bed) soon.

Slowly moving further along the lakeside I soon gained a good view of a large cliff which loomed westwards over the slopes above the southern end of the water's edge. It was a true glacial landform: smooth and gently sloping on its eastern, iceward side; sheer, rocky and viscerally weather-beaten on its western side: an A-symmetry consistent with the bygone action of ancient ice. Meltwater beneath the historical ice sheet seeped into rocky cracks and fissures on the down-stream side of the mountain, only to re-freeze and expand, quarrying colossal volumes of rock out of this ancient surface. Concordantly, at the base of this cliff there lay a number of large boulders, and by large I mean truck-sized – the kind you can crawl underneath and feel almost at home. This was a perfect place to bivvi. The cliff would shield me from the ravenously cold easterly winds which batter down from the high pressure zone over the ice sheet, and these boulders would provide an extra line of defence against any surprise rain showers which might conspire to 'enhance' my feeling of exhilarating exposure. So, I laid out my sleeping bag, stuffed it into my camo-green Royal Air Force issued bivvi sack, stuck my boots toe-to-toe by the open end in lieu of any form of pillow and hunkered down for a nice, contemplative evening in the chill Arctic air.

It was clear and crisp, and almost perfectly silent. The only sounds came from what I'm pretty certain was a family of Greenlandic gyrfalcons nesting high on the cliff above my head. Gyrfalcons are beautiful birds, sporting a striking white plumage into the winter months. But, to be honest, their shriek isn't the most relaxing of nature's lilts. Although I'm a bit of a closet birder (there are people out there who will never let me forget having made that statement), and I'm always thrilled to see unusual species in unusual places, tucked up in my bivvi in this almost pristine amphitheatre of rock, sky and tundra, the periodic sound daggers coming from the crags over my head had the effect of somewhat upsetting my calm. Nothing arouses feelings of being hunted, or stimulates half-formed realisations of one's inevitable mortality, quite like the roar of a predator or the shrill call of a distant bird of prey, especially when alone, exposed and cut off from any of life's usual comforts.

It was all pretty atmospheric. I loved it.

The rain held back, the clouds took the evening off, and as luck would have it that night I was treated to unquestionably the best display of aurora borealis I have ever seen. In fact it was perhaps a little too good as I failed to make even an attempt at sleeping until well past two in the morning. Instead I lay on my back – wrapped in my bivvi but in all other ways completely open to the wildness of my environment – and simply watched.

The air was cold and dry, the rocks were magnificent in their shadowy motionlessness, and all the while above me the sky sang its silent spectral song. I will never forget it. From the point at which darkness began to overwhelm the now departing day, ethereal wisps of blue light started encroaching on the horizon. As the last vestiges of the sun's influence faded from the heavens those blues transformed slowly into vibrant greens, and soon the whole sky was ablaze with writhing, verdant flames. The sheets of colour moved together like kelp fronds swaying on underwater currents; so dynamic, but yet almost invisibly. It was overwhelming. To sleep would have been almost disrespectful, so on and on I watched. At one stage, approaching one o'clock in the morning, reds and oranges began to appear amidst the greens and if I looked directly up, towards the centre of the sky, I could see what I can only describe as the 'crown' of the lights: the central point from which all the thrashing colours seemed to cascade. It was almost as if there was a hole in the sky, and through it, all the colours and splendour of another world were spilling.

For all those who have never seen the aurora, you must. It's unforgettable, and singularly moving. Those beams of light seem so uncannily alive it's almost disturbing, so much so that on the Isle of Lewis, in Scotland, people refer to them in Gaelic as 'the tall men', so reminiscent they are of spirits moving amongst each other. In fact, the first time I ever saw the aurora was in Scotland, aged circa five years. It was a long time ago but I still remember the sensation very clearly. I was terrified! All I knew back then was that there was something VERY wrong with the night sky, and I didn't like it one bit. I'd seen plenty of night skies by the age of five: black ones; grey ones; starry ones; rainy ones. But never a night sky that moved! That wasn't OK as far as I was concerned. Darkness is scary. Every kid knows that. But at least darkness doesn't move! That means it can't run and catch you! Well, that's what I'd thought up until that night anyway. I ran into the house and didn't leave it until morning.

Fortunately, by the time I was twenty-seven I'd seen the aurora a few more times and had come to accept that it wasn't some daemonic display of supernatural death-light. It was simply beautiful. And so, lying there beneath those painted Greenlandic heavens, the waterfalls of light which rent the sky eventually washed the vitality from me and I was coaxed to sleep, at least for a few hours. In what seemed like a painfully short time I was awake again, only this time with the dawn. I opened my eyes to find myself lying in the midst of a freshly sprouted miniature jungle of tiny, fragile frost crystals. It was a beautiful sight to welcome me into the new day, although perhaps not amongst those most likely to make me actually want to get out of my soft, warm sleeping bag. But rise I did, and after a bit of time and effort spent wrestling my now frozen hiking books onto my quickly cooling feet, I packed my minimalistic kit away and began walking.

Over the next couple of days I wandered amongst lakes and hills, passing herds of musk oxen and the occasional caribou along the way. There was not another soul to be seen. It was wonderful to be alone and free in this landscape after so long spent killing time in Kangerlussuaq, and for the first time since stepping off the plane almost two weeks before I genuinely felt like I was in the Arctic. I hiked over bog, grass and heather all the way to the expansive fluvial plains which emerge from the edge of the mighty Greenland ice sheet, and here I stopped. From a vantage point on a hillock overlooking the sandy, braided river network below I marvelled at the way in which the familiar rolling hills of west Greenland – so similar in many ways to the views I grew up with in the west of Scotland – are suddenly subsumed by this other-worldly mass of ice which flows perpetually from the eastern horizon. Here was where two worlds met; the shores of Greenland's ice ocean.

In many high-latitude or high-altitude environments large ice masses terminate as channelled glaciers which extrude through mountainous topography like great frigid fingers, each finger a great buzz-saw, cutting down through rock and clawing at the land as they gradually melt. In the more southerly parts of west Greenland, however, those glaciers are shorter and broader, and from where I stood it seemed that I could see the whole of the ice sheet. The mountains are lower and the landscape more open, a tribute to the relatively short amount of time which has elapsed since the still-emerging bare land was itself buried beneath the adjacent expanse of ice. This was not the maritime, fjord-punctuated ice edge of farther north; it was more like a horizon. It was a stunning view, and one which felt singularly poignant to me at that time. Here I was, at the edge of my habitat; where rocks, soil and life meet the cold, sterile ice of the interior. And soon I would be at another meeting place of worlds; at the shores of the ocean. And there I would set foot from one life into another, from a life on land to a life on water. And what would happen then? That was yet to be seen.

I trekked back to Kangerlussuaq in something of a melancholy mood. The small blocks of cheese, salami and bread which I had brought along to keep myself fed whilst hiking (remember I had come packed to sail, not to camp) were now finished and my time on land was drawing short. Very soon 'tomorrow' would become today, and it'd be off into the unknown for Matt, Sam and myself. But there was no sense worrying. I knew I wasn't going to bail on this one. It was very much a do or die decision that I'd made when I agreed to undertake this impending voyage. Well, perhaps it was more of a do AND die decision! I smiled at that thought. It really didn't matter. I was sailing, and that was all there was to it.

6 / Reporting for duty

A couple of days after I returned to Kangerlussuaq I found myself boarding a small twin-engine Air Greenland turboprop 'Dash' aircraft. Alun had appeared in Kanger at about the same time as I had emerged from the tundra, and now was at the airport to see me off. He passed me a couple of extra bags containing kit for the boat which, up until now, had been in our Kangerlussuaq lock-up, shook my hand and, worryingly, even gave me a hug! Surely this couldn't bode well! I don't remember all of what he said to me before I got on the plane, but I do remember his last piece of advice:

"Don't worry. Sailing across oceans is easy! There's nothing to crash into out there!"

And, he was absolutely right! There is almost nothing to hit! To be honest though, it really wasn't collisions I was concerned about at this stage. No matter, however, as they were true words nonetheless. And it was time for me to go. I walked out of the departures room, crossed the tarmac of the flight line and, without a backwards glance, climbed up into the red flying tube bound for the coastal town of Sisimiut. It was here in Sisimiut that I'd be meeting with Gambo. Matt and Sam were already aboard, having sailed down from the more northerly town of Uummannaq where they had both recently been struck by the meteorite of spontaneity and opted to join me in the salty, adventure-laced marinade of the North Atlantic.

This was it! I was on my way!

A mere hour later the Dash touched down in Sisimiut. We taxied up to the small terminal building and the engines began to whine, winding down to a standstill. I stood, gathered my meagre belongings and slotted into the small but typically impatient crowd of people who were eager to disembark. Honestly, I can't exactly say why I was so keen to leave the plane, but nonetheless I found myself jostling with the best of them as we all slowly waddled towards the exit hatch.

I reclaimed my bag and began walking into town. This wasn't actually my first time in Sisimiut. I had been here during my first visit to Greenland, seven years previously in 2003. I had flown in on that occasion as well and, being something of a tight-fisted Scot when it comes to paying for public transport when walking will suffice, I had opted to ride the old leather, personnel carriers (i.e. my boots) into town back then as well. Now this is where my memory fails me, for on one of these occasions I was actually picked up by a friendly Danish chap and given a lift into town in his pick-up truck. (I have no idea whether this happened in 2003 or in 2010 however.) So, either it was a long and sweaty two-mile walk into town carrying several large bags, or I was very grateful for the lift, what with being so encumbered by the extra load Alun had kindly furnished me with. One way or the other, I remember descending the road through Sisimiut's small, town centre and approaching the harbour area. Here, in 2003, I had caught a small overnight ferry north to Ilulissat (a ferry which, incidentally, later found a new lease of life in the Polar Regions cruise industry – in which I currently work – under the new name *Ocean Nova*). This time however I was looking for a much more diminutive vessel.

And there she was: Gambo; the only yacht in the harbour. It was like seeing an old friend again.

7 / An inauspicious start

I boarded Gambo to find the atmosphere somewhat subdued. This was unfortunate for at this point adrenaline was pumping through my system and I was being anything but quiet. Gambo and her current crew had only just arrived a few hours previously and weren't in the best of spirits. It had been a pretty choppy passage from Uummannaq and everyone was tired. Aboard had been Nolwenn, Max, Sam, Matt, an Edinburgh university PhD student named Tom Cowton and a Swansea University PhD student named Christine Dow. Christine had just left to catch the next flight to Sisimiut. Max, sporting a bandaged hand which he had acquired some weeks previously and which was preventing him from remaining aboard, was in the process of packing his gear and would be leaving as soon as possible. Matt, Sam and Tom were helping Nolwenn get the boat ready for the next leg of her trip and were pleased to see me, but Nolwenn was very much preoccupied (and I suspect pretty worried about imminently leaving us in the driving seat of his charge). Sam, although positive, was very under the weather having been badly ill, seasick, or both (it's hard to say) over the course of the last few days.

A few years later, as I was looking back at this whole adventure, I asked Sam about this initial trip down to Sisimiut. These are just a handful of the snippets which stood out:

I was stuck, trying to sleep, in the bow of the boat getting 'good air' as she went through the waves.

I don't really have any memory of Matt being there at all, but he must've been.

Things (heavy things) kept on falling on me.

Matt was on pretty good form, but, as I would soon learn, Matt usually is. This was actually the first time I had ever met Matt in person and I was instantly impressed by his confidence, energy and enthusiasm. Yet even he however was a little less bubbly than I would soon come to expect as his 'usual'. I think the reality of what we were all preparing to do was beginning to set in, especially given the rough weather they had just had and also the fact that now people were actually 'bailing out'. Christine had left, Max was about to go and now they were turning around as quickly as possible so as to get us all to Nuuk where Tom and Nolwenn would depart. It was all a little full-on, and I had descended right into the middle of it. I think to those crew already aboard my arrival must've felt a little like an excited child exploding into its parents' bedroom at 6 am on Christmas morning.

Regardless, I was able to say a brief but warm hello to Max, whom I had sailed with in Canada several months previously. It was great to see him, and I was gutted he wouldn't be with us over the coming weeks. But *'the best laid plans of mice and men....'* Max's hand was a show stopper for him and he somewhat enthusiastically departed less than an hour after I crash landed on the scene, bound for France, home and his much missed girlfriend. I was disappointed, but, there wasn't to be much room to dwell on these feelings as within an hour, and almost before I'd even had a chance to claim a berth and stow my kit, Nolwenn slipped our lines and we were off, back out into the Davis Strait and bound for Nuuk – Greenland's capital.

This was it! I was at sea again! It was all a bit of a whirlwind and I wish

I had more to say about this first foray, but any feelings I must've had are obscured partially by the speed with which this all happened and also by the fact that within two hours of putting to sea I was stricken with seasickness. Wonderful.

Sam and Matt had retired to bed, eager to claw back some energy after their arduous passage to Sisimiut, leaving Nolwenn and Tom on watch. I stayed up on deck, keen both to catch up with Nolwenn and also to get to know Tom, who straight away seemed like a really good bloke. So, we chatted away for a while, listening to Nolwenn tell us how one reliably predicts a coming storm from the visible cloud formations and also exactly how one should correctly go about changing the transmission in a Yamaha outboard motor.

Soon Tom felt compelled to go below and make some dinner, and so before long the sickly sweet aromas of chicken curry began to creep like invisible tendrils up the companionway, through the hatch and into my rapidly destabilising stomach. But, I muscled on. We continued to chat, and when Tom produced a plate of piping hot fodder I gladly took it from him and enthusiastically wolfed it down as only a young man eager to prove how tough his stomach is can. It tasted great, at least it did the first time it passed through my mouth. About twenty minutes later, as it was making its second passage over my tongue, the flavour had changed somewhat. So, after getting my first good look over Gambo's side and generously feeding the fish I succumbed to the really rather modest waves and poured myself down into the cabin and into my bunk. It's fair to say that at this point I was asking myself some fairly pertinent questions about what the hell I was doing here, just as I expect Matt and Sam had asked themselves a couple of days previously. But, onwards and upwards! Or was it upwards, then downwards, then upwards, and downwards, then upwards, mixed in with a good bit of side to side and back and forth.... Urgh.

Fortunately the distance from Sisimiut to Nuuk isn't huge, and by the time I was roused from my sick bed we had completed the lion's share of the passage. In all, it took us less than twenty-four hours and, in the darkness of the very early morning I stood with Nolwenn on the stern of Gambo, spotting for the all-important yellow leading lights which would guide us through the scattered islets and channels which guard the entrance to Nuuk harbour. By this point I felt like a new man. Eating, de-eating and then sleeping had apparently re-set my system and I had quickly found my sea legs. In fact that would prove to be my only 'fuel dump' of the whole adventure. Truly, there would be times when my cargo bay doors felt sure to burst, but from this point onwards freight went in where it ought to and left the building by the conventional exit, praise be.

And so, Gambo slipped under the cover of darkness into the busy port of Nuuk, surrounded by a motley assemblage of trawlers, commercial shipping vessels, research craft and a handful of other yachts. This was the end of the road for Tom and Nolwenn, but for Matt, Sam and I, this is where it would all begin in earnest. So, after coming alongside a large fishing boat and securing our lines to her deck Nolwenn brought the engine's spluttering protestations to an abrupt end and we all retired once again to our sleeping bags. From these already moderately salty sacks we would await the coming dawn and the changes it would bring.

8 / Letting go of land

Nolwenn and Tom stayed with us for three days after our arrival in Nuuk. This gave us time as a group to: properly tidy the boat; shop for enough food and morale (by 'morale' I mean Haribo jelly sweets) to see three hungry blokes through potentially as much as a month at sea; stash everything in a shipshape manner; refuel; fill the freshwater tanks; and properly service the engine. The 'magic' propulsion box got new belts, fresh oil, a new air filter, and had her rocker cover whacked off so we could check the tappet spacings. It was a proper job, although I'll admit to having felt a little uneasy about seeing our engine in bits so soon before our departure into the unknown. When there is a big job on the horizon, and success stands to depend heavily on a particular piece of equipment (i.e. in this case the engine) I'm often liable to avoid touching it before the hour of need, on the grounds that "if it ain't broke, don't fix it!" Still, in this case, I was glad to make the most of the experts we still had at our disposal. I'm pretty sure Nolwenn felt a little bad for having to abandon us, as he really put in the hours to set us up before he and Tom caught their Iceland Air flight out to Reykjavik. (Either that or he simply didn't rate our chances too highly, and so wanted to give us the best start possible!) Either way, we were really glad to have him there to help, and felt bloody awful when finally the time came for him, and Tom, to go.

We did make last ditch efforts to persuade Tom that in fact sailing home was a far more sensible option than simply flying, but he was having none of it! Several hours in comfort and relative safety seemed to appeal somewhat more to him than upwards of two weeks of fear, intense discomfort, perpetual wetness and intense psychological strain. Even we had to admit, his reasoning was fairly bulletproof. I think we had just wanted to see if he would even consider it. If he had done, or if he had seemed even a little tempted, perhaps we could have persuaded ourselves that in fact we were NOT complete fools, and that this wasn't at all an insane venture for three novices to take on.

In fairness, I think Nolwenn genuinely did want to stay, but he really couldn't. The real world was calling and he had no choice, but we weren't too happy about it, although we made sure not to show it too much. When he left that would be it. No-one to ask for advice; no-one to take the helm in a fix; no-one else to blame. There was no sense dwelling on it though. We just got on with things and enjoyed the extra hands and good banter while we could.

Shopping was great fun, especially as it was ultimately on Alun's expense and not our own. We bought an impressive quantity of tuck. When you work it out and buy it in bulk, it's pretty frightening how much food three adults need to see them healthily through the best part of a month! All in all we spent several hundred pounds, or rather Sam spent several hundred pounds, which Alun later paid him back. Plenty of sensible items were bought, things such as pasta, flour, eggs, oranges and plenty of instant noodles (these would prove to be a godsend). Of most importance though were the high energy, high morale, low-effort items. These were things like chocolate biscuits, cured meat, cheese and the Haribo sweets. I knew from experience, and the other lads knew from common sense, that when the world is constantly moving and you are perpetually tired and wet, the last place you want to be is below deck hunkered over a cooker. Fast, effective and tasty foodstuffs were top of our list, as were fresh vegetables and fruit.

We knew these wouldn't last the whole voyage, but they would improve our world no end while they did.

Water is also crucial. You do not want to run low on fresh water whilst at sea. I had been in this situation once before whilst sailing in the Antarctic, where we had to make roughly eight gallons last eight grown men nearly a week in the Southern Ocean. That will forever be locked in my memory as probably the lowest physiological (and thus also psychological) point in my life. It's not something I like to think about, but I did learn from it and I always make damn sure I have plenty of water when I go to sea now.

Gambo has several large water tanks, and we needed a fair bit to fill her up. However, none of us were sure where the water point in the harbour was. So, I went exploring while the others got stuck in sorting and storing our fresh provisions. I was glad to escape this job because in all honesty 'tidying up' has never really inspired me. Instead I headed over to where the bigger ships were and began asking around about where the hose was to be found. And they really were BIGGER ships! At this point there was a large Danish cargo ship alongside which probably measured in excess of 120 metres. There were also several large fishing trawlers registered in Newfoundland and one truly massive cruise ship aptly named *The World.* I avoided the cruise ship on instinct and asked the various bods who were milling around the more utilitarian side of the dock. I actually had a surprising amount of trouble locating the water supply, but eventually spoke to a chap from the cargo vessel who said he would find out for me. This involved being invited on board his ship, which I was pretty pleased about. He took me up to the bridge and offered me a cup of coffee (which I naturally accepted, especially as it was unlikely to taste of engine oil, which, to varying extents, pretty much everything on Gambo did).

This bridge officer (I have no idea what his name or rank was) made a number of phone calls, presumably to the bosun or to the engine room, and eventually someone appeared on the bridge who could show me where to find the water supply ashore. I necked the remainder of my coffee, thanked the bridge officer, stepped outside and clambered down a rusting steel ladder after my new guide. We disembarked via an aluminium gangway and walked back along the quayside to a somewhat makeshift looking wooden box which had been erected beside one of the large floodlights which lit the harbour during hours of darkness which, of course, vary hugely throughout the year in this part of the world.

After a few moments of wrestling with the latch mechanism my guide succeeded in opening the box, revealing a long coil of hosepipe which was open at one end and was attached to a tap at the other. Result!

My guide gestured at the hose, nodded, and went on his way so abruptly that I almost didn't have time to thank him. I don't think this abruptness was intended to be rude, or even necessarily a sign of any resentment, as the bloke smiled broadly when I caught his hand in thanks. I got the impression that his matter-of-fact demeanour was more habit than expression, perhaps the product of spending many long weeks and months of his life at sea performing the same repetitive tasks in the same space and in the company of the same people. Interesting.

Anyway, I returned to Gambo and we motored over to the water point. After tying up alongside a large fishing boat from Newfoundland (or Newfie, as the crew called home), we proceeded to clumsily fill our tanks, spilling a huge quantity of water in the process as the hose was too large to fit into our tank's fill point. This reminded me of crewing Gambo in the Falkland Islands many years previously, where we had been taking on a British Army charter group at Mount Pleasant airbase. The British military were supplying us with provisions and the like before we sailed for Antarctica (from whence we had just come), and when it came time to fill the diesel tanks, had sent a large fuel tanker truck, or 'bowser', down to Mare Harbour where we were moored up. I'll always remember the look on the operating Sergeant's face when he and the young private who was accompanying him unhooked the trucks outlet pipe, which had a diameter of about four inches, and approached our fuel inlet, which had a diameter of about one-and-a-half inches. Hmmmm.... What ensued was a lengthy and fairly comical operation of (very carefully) pouring diesel from a large British Army fuel truck into an everyday bucket, and then transferring it via a large funnel into our tank. Only on Gambo!

And so, slowly but surely, we tied up all the loose ends that needed our attention in Nuuk. Our inevitable departure was getting close. I have no recollection of what day it was when Nolwenn and Tom finally abandoned us. Days of the week very quickly cease to have any relevance at all when you are sailing, unless of course you need to be aware of them for weather forecasts or tidal information, but only then when you are in the coastal, day-hopping game really. Our wee trip was set to be a single leg 'no stop wonder', and we needed to leave soon as autumn was already in full swing.

The day before our departure I set off, armed with our passports, to find the local police station and 'clear customs'. Meanwhile, Nolwenn took Gambo and the lads over to another part of the harbour where the fuel pumps were in order to fill our tanks with diesel, arming us with seven days' worth of 'go go juice'.

As I skulked around Nuuk in search of the police headquarters I quietly ruminated on why I felt it necessary to announce our departure at all. I really didn't think that anyone would mind if we were to just slip out unannounced. No-one had taken any interest when we left Canada the previous April, and here we were essentially in the European Union, destined for another EU port.... Did I *need* to clear customs? No, probably not is the answer, but for some reason I felt the need for a bit of a ceremony over this particular departure. Firstly, I wanted a passport stamp to look at in years to come, to remind myself of this little phase of my life. Secondly, and perhaps most pressingly, if things went badly, i.e. if we never made it back to the UK (I have a vivid imagination), I wanted there to be some record of us having left. You know, like in the movies.... When the Hollywood actor as private investigator goes sniffing around in an attempt to piece together the final days of our short lives (for some reason I'm imagining Russell Crowe). I wanted there to be someone in the Nuuk police station who could say:

"Oh yeah, I remember that guy! A bit skinny. Kinda nervous looking. Smelt like mould.... He was in here alright!"

Something like that anyway.

So, I found the police station, and after only a little persuasion the guy on duty, who was very friendly and perhaps more than just a little bored, agreed to take down our details and rummage around for the emigration customs stamp. And so I achieved three passport marks, which I have to say were a little disappointing really! I was hoping for something Greenlandic with at least a whiff of the exotic. Maybe a polar bear? In the event, we got the standard EU stamp with "DK" for Denmark in the upper left corner, Kalaallit Nunaat as the port of departure, and a little outline of a ship in the top right to denote the means of our egress. Actually, I was quite excited about the little picture of the ship. It felt…boaty. Simple things.

By the time I got back to the harbour Gambo was just coming alongside once again, tying up as before onto a large fishing boat. I received the lines and secured them, allowing the lads on board to concentrate on fending off. Nolwenn turned the key anti-clockwise and the engine clunked to a halt. That was it. His job was now done. We sat down to enjoy a somewhat solemn cup of tea while Matt engaged in some frantic note taking. Matt had been writing down some of the hot tips and snippets of advice that Nolwenn had been feeding us these last few days, and had compiled these gems into a small volume which he entitled:

"How to sail across the Atlantic; an idiot's guide, by Nolwenn Chauche – possibly the best in the world at everything, ever".

I believe this volume still exists.

Eventually, Tom and our skipper said their farewells and left to catch their flight. I think at this point the three of us who remained drank some beer, probably threw some typically British dark humour around, and went to bed. It's a good thing we'd had some booze, as I likely wouldn't have slept otherwise!

That night the aurora danced again, painting green snakes in the eastern sky. I spotted them as I crept up on deck to 'empty my ballast tanks' under cover of darkness, and managed to get a few blurry photos in. Blurry that is because it's a little tricky to pull off nice, crisp long exposures of the night sky from a floating platform! In fact, there are many traditionally simple tasks which can be difficult to perform on a constantly moving platform where gravity can never be relied upon! Blurry photos were just the start; a taste of what was to come.

The next morning we were awoken by a tapping sound on the hull. We poked our still foggy heads through the companionway and encountered an Inuit fisherman. He was from our neighbouring vessel; the one we had tied up alongside the previous day. Their engines were running and they wanted to get out. We would need to recover our lines and move off to let them go.

We hurriedly dressed, threw some boots on, and got ourselves up on deck. I checked the engine was in neutral, counted to ten as I warmed her glow plugs and fired her up. At that point Matt and Sam both looked at me. This was a bit of a shock at first, but then I remembered that I was the only one who had ever really helmed Gambo before, and I was ostensibly supposed to be the skipper now, ridiculous as we all knew this was. The situation felt pretty odd, but I had done plenty of mooring and berthing with the lifeboat back home in Aberystwyth (I had, at this point in my life, been a crewman on Aberystwyth Lifeboat for almost three years), and I had spent a lot of time on Gambo with Nolwenn and Max in Nova Scotia that previous spring, coming in and out of a lot of ports. So, I just did what I thought Nolwenn would have done.

"OK guys, could you remove the springs first please. Sam, could you take the stern lines, and Matt, you take the bow lines."

We loosed the springs, brought them on board, and then I asked Sam to undo and recover the stern line. Meanwhile Matt set the bow rope up as a slip line which could be held fast from our end but released and pulled through when necessary, without the need for someone to untie it on the other boat. This done, Sam pushed us off at the stern. As we gained a little clearance I put Gambo in reverse gear and gave her a little throttle, steering directly aft. As it happened she didn't steer directly aft. Instead she pulled slightly to starboard. I would soon come to appreciate that yachts with rudders are harder to manoeuvre in reverse than powerboats, (a lesson that I would later learn more fully in Alderney Harbour in the Channel Islands when, in 2012, I managed to almost jam another yacht directly across a small harbour entrance). For now though, there was no issue. The only effect was that the bow moved outwards slightly as well, which was actually a benefit. Matt let the line slip and after we had sufficient clearance I put Gambo in forward gear, brought the helm about and steered slowly away into the harbour. The local Inuit lads on the fishing vessel waved their thanks and began going through a similar process, untying from yet another fishing vessel which was even larger still. They freed themselves and moved off, heading out of the harbour altogether and turning off to the west, out of the fjord system and away in the direction of the Davis Straits and the open sea. I began to bring Gambo back into a position to come alongside once again.

At that point we had something of 'a moment'. All three of us exchanged looks. For a few seconds we were silent, and I brought Gambo to a standstill, giving her a little power in reverse and then popping her in neutral.

"Is there any point in mooring up again?"

It was Sam who was asking this question. I think Matt and I just sort of shrugged. I probably exhaled sharply through my nose, in a kind of half laugh, and said something like: "Nope, I suppose not! It's a nice day. We've done everything we need to do. If we tie up now it'll just be for the sake of untying again in a couple of hours."

"Let's do it!" said Matt.

And so we did. I put Gambo back into forward gear, slowly turned her through 180 degrees and steamed away in the tracks of the fishing boat which we had only just let out. It appeared that this was it. The game was on!

9 / To sea!

14/09/10
1230 UTC. Sam writes in log:

Depart Nuuk towards Oban. Engine start @ 1230Z; Matt, Sam + Colin on board. Full fuel + water. Dist to dest (DTD) = 1861 NM.

The weather on our first day out of Nuuk was spectacular. The sun shone brightly down from a nearly cloudless sky and the waters cast dappled reflections up at us. There was barely a breath of wind. This was, perhaps somewhat counter-intuitively, great for morale as it meant we could forestall any actual contact with the sails. I for one have always been in fearful awe of the wind and its sometimes bewildering strength, just as I have always felt deeply terror-laced admiration for the sea. Thus, despite having willingly embarked on an epic sailing adventure, I was in no rush whatsoever to get stuck in about the canvas! We were in utterly no doubt that there would be more than ample opportunity to fill our boots with good old-fashioned wind wrangling over the coming weeks and so here, in the fjord, amongst islands, tidal currents and a smattering of other vessels, we were more than happy to rely on the engine. We had a full seven-days-worth of fuel after all, so surely indulging in a good dose of motor cruising at this stage couldn't hurt!

The first other vessel we passed as we slipped out of the harbour entrance was, of course, the large cruise ship which was still in town – The World. Leaving the harbour we puttered past almost literally beneath her bow before then turning westwards and heading off down along her starboard side. We passed a few of this big white leviathan's crewmen who, hanging on wooden platforms suspended on ropes from her upper decks, were repainting patches on her hull. As our modest wake lapped against the waterline of this colossal seagoing city the irony occurred to us that we had been under-way for less than five minutes, and already we had sailed half-way around The World! Ha. Job done! We might as well go home!

'Home', at this stage, seemed very far away indeed. Shortly after leaving Nuuk I took the time to input a few key waypoints into the GPS (only a handful really, as there aren't all that many obstacles, landmarks or hazards in the North Atlantic). The waypoint entitled "Oban" (this being our final destination) was precisely 1858 nautical miles away. Looking at that number on the GPS display kindled several conflicting feelings inside me. I felt despair at the enormity of the task ahead; excitement at the magnitude of the adventure; fear at the unknown and of the open ocean, and also a measure of pride at having chosen to undertake the whole enterprise. I don't think I've ever felt such tremendous exposure in all my life as at that point, as we motored out of Nuuk in blazing sunshine and with absolutely no intention of turning back.

I suppose I can only speak for myself on this one, but I was scared. Very scared. I made every effort not to show it though. Of course, being three lads in our twenties, we predictably kept any feelings of trepidation or fear pretty close to our chests. This was made easier by the high pressure on the barometer, the pleasing greens and blues on the weather chart we had just downloaded over the satellite phone, the playful sun which beat down upon us and the lethargic waves which Gambo effortlessly ambled through as we steamed towards open water. Despite any fears our spirits were high.

We were all smiling. It was great. I wondered if maybe we were invincible after all! Matt in particular was in great fettle, infectiously so. His positivity and almost tangible joy at simply undertaking an adventure such as this was to be a thread throughout the whole voyage. I swear that if at any point we had started to sink there is a real possibility that Matt's buoyancy alone would have been enough to keep Gambo's full 15-metre length of sturdy steel afloat.

We passed our first iceberg almost immediately, still close enough to Nuuk that we could have made out faces in the windows of the shore-side houses. Its above-water portion was only about the size of a family car, but an iceberg it was nonetheless. Naturally we took full advantage of the moment and rattled off a number of comically posed pictures of each other at the helm as the ice slipped past. We did our best to look daring, glamorous, intrepid and just downright badass. Just as well really, as that would be our only close pass to a berg of any size!

Ice was, of course, something we were a little nervous about. Wholly amateur we might have been, but idiots we were not. Two of us were glaciologists after all, and I had logged more of my sailing time in the Polar Regions than anywhere else. So, we were ready for the bergs, at least psychologically. Given the now failing light which comes with the approach of winter we knew that the navigational hazard presented by errant ice cubes was something we would need to take seriously. Now, I'm sure that ice charts and ice forecasts would have been available via our feeble satellite phone's internet connection if we had ever thought to look for them, but being little more than sailing greenbacks this idea, much like an imagined nightmare iceberg, never even came onto our radar. We just weren't 'plugged in' to the bigger picture of maritime information networks. Later experiences of working on larger ships in the Polar Regions has since opened my eyes to the amount of information which is available, but in 2010 I just didn't know where my support network was, or that it even existed for that matter. I thought maybe the best way to get information would be to cross the playground and ask advice from the 'bigger boys'. So when, whilst entering open water, we passed a large, orange, utilitarian-looking vessel with Copenhagen marked on its hull as port of registry I made note of the ship's name and jumped on the VHF for a chat. This made me VERY excited. I love ships. Vessels to me are like people. They all have names, go on amazing adventures to far-flung parts of the world and see places, phenomena and 'character-building' weather conditions that most people on Earth never will. Generally speaking, the bigger the ship the wilder her adventures and farther flung her travels. Thus, from my perspective at the time, having only ever worked on a small sailing boat, big ships were (and to a great extent still are) a bit like celebrities. Naval vessels are like action heroes and Arctic or Antarctic supply or utility ships are like famous explorers: The Ernest Shackletons and James Clark Ross's of this world. Both of these names have now deservingly been attributed to British Antarctic Survey support ships and in the years following the events which I recount in this book I encountered both these and numerous other high-latitude working ships (such as the US supply ship *Lawrence M. Gould*) while I worked on polar expeditionary cruise vessels. Sighting them always gave me a similar buzz to that which you might feel when spotting a celebrity in the supermarket. I would always get excited, and generally get on the public announcement system to let the passengers know that we were passing a local maritime personality. It was always interesting to me how passengers would generally share my enthusiasm. It seems that ships really do occupy a strangely special place in many people's hearts, and not exclusively in the hearts of people with seafaring experience and associations. There is perhaps something in the adventurism of an ocean-faring ship, in the fact that ships have names, and perhaps also in the reality that ships, like people, have a working lifespan that draws us to them almost as if they were people themselves.

Anyway, calling this larger vessel on the VHF was a bit of a thrill. Still, having never been one to shy from either a challenge or from proactivity, I got on the horn and called them up in search of some advice on ice conditions. And they answered!

To be fair, it was a bit silly of me to assume that simply because the ship was registered in Copenhagen this meant that she had just come from that direction, but as luck would have it she had in fact just rounded the southern tip of Greenland in the last couple of days and was able to share her observations on ice conditions, which was nice. Our concerns were rooted in the fact that cold water currents originating in the higher latitudes sweep down Greenland's eastern coast, carrying large icebergs southward from the tidewater glaciers which calve in that area. Some of the bigger bergs survive long enough to kick around Cape Farewell and thus end up in the generally more ice-free waters of Greenland's south-western coast, where the inland ice frontier is many miles from the shoreline. These were the waters which we would be passing through. Titanic territory!

Fortunately the news was good. Although visibility had been poor when they had rounded Cape Farewell, the bridge officer to whom I spoke described virtually ice free waters. Of course the radar which they carried was much more powerful than ours, so poor visibility wasn't as much of an impediment to them as it might be to us. Still, they hadn't picked up any bergs within range of their instrumentation. They did however mention that the weather was deteriorating and that in the time it would take us to cross to the other side of 'the pond' we were likely to hit some 'discomfort'. We already knew this, or at least we already had assumed this would be the case, and were committed to the voyage regardless, but hearing it from another much larger vessel which had just passed through the same waters for which we were bound didn't help morale much. But at least the ice outlook was good! Fortunately this VHF exchange ended on a high note as it turned out that the ship was a survey vessel and was carrying some researchers from the UK. When we stated that our next port was to be Oban one of them overheard and, apparently being from the Isle of Mull (on Scotland's west coast) herself, she came onto the radio to wish us well. This kind of sentiment is always appreciated, especially at sea, and so we reciprocated in typical laddish fashion by inviting her along to the party which we planned to have upon our eventual arrival in Scotland. Perhaps it was a touch optimistic, or maybe naïve of us to think that we would be in any state to make good on those plans, but that was the future. In the here and now we still had the lion's share of 2000 miles of open water to contend with.

10 / Finding a rhythm

All fears for the rest of the voyage aside, this first day was simply bliss. The mountains in the east slowly dwindled away and our nerves began, almost imperceptibly, to melt away amongst the gentle vibrations emanating from the engine compartment. Sam indulged in a spot of fully-clothed sun bathing and I took the opportunity to get a few mugshots of my crewmates about the deck. It was wonderful, and disarmingly easy.

It's frightening how rapidly lethargy can set-in during situations like that, especially when your gut knows that ultimately there is acute discomfort somewhere on the horizon. Part of your brain seems to recoil into a ball, drawing a lot of your creativity back behind the mind's parapet. I think the first sign of this was that we eventually grew weary of the novelty which manual steering offered and opted to engage the boat's autopilot. To operate it one simply pointed the boat in the direction you wanted to go, held that course for a few seconds, and then pressed "on" on the main navigation control panel. The autopilot then communicated with the GPS to ensure that the boat held this course, forcing the tiller from side to side with a hydraulic arm. This hydraulic arm was located in the port-side stern locker and was driven by a servo motor situated inside the main cabin on the wall of the "skipper's berth", immediately to port of the companionway. This servo was damn loud, especially if you were the one trying to sleep in the skipper's berth with your head next to it, but we would eventually come to find that this noise was somehow a comfort. On occasions later in the voyage when it stopped without prompting the silence was deafening, as this silence generally meant that something had broken and we were literally out of control. We grew so attached to the sound of the autopilot, and so fond of its contribution to our life at sea, that we gave it a name. We called it 'Geoffrey'.

Geoffrey really was very much part of our little family. He had a moustache and exotic-looking eyebrows which we cut out of decking foam and glued to the small semi-spherical GPS receiver which was mounted on the cockpit dodger, just above the main companionway hatch. This little grey blob was our fourth crew member, and by far and away the most robust of us all. He was always on watch. He stayed topside even in the worst storms and, although occasionally fiercer seas would put the wind up him, he never lost his rag or made a negative comment about anything or anyone.

Eventually, after a couple of hours spent dawdling out of the fjord, we (well, Geoffrey actually) made a turn southwards, assuming the course which we would follow for the next two days and which would take us down the south-west coast of Greenland, first towards the rather beautifully named Cape Desolation and then eventually sufficiently far south to make our eastward turn around Cape Farewell (which in addition to being an iceberg slingshot zone was also clearly named for morale purposes). It was at about this time that I approached the boat's surprisingly well stocked library in search of a good novel. Something to keep me distracted over the coming weeks. There were plenty to choose from. The titles were pretty varied, ranging from a dog-eared copy of *The Lord of The Rings* to Karl Marx's *Das Capital*; and *Storm Survival Tactics*, a Lonely Planet guide to Brazil, and *Lord of the Flies*. A real mixture of adventure books, technical manuals for climbing or sailing, philosophical or classical literature and a smattering of good old-fashioned travel accounts; all pretty representative of the types who generally accounted for the majority of Gambo's standing crew base.

For some reason, in the face of this wealth of options, I selected *All Quiet on The Western Front* by Erich Maria Remarque. Why I picked this particular book I honestly don't know. Perhaps I wanted to remind myself that no matter how scared I got over the coming days and weeks, things could always be worse!

Written from a German perspective, *All Quiet on the Western Front* offers a painfully detailed and candid account of life and combat in the front-line trenches of the First World War. When I started reading it I was quite literally coming to terms with the belief that I would likely drown at some point during the next fortnight or so. A visceral portrait of trench warfare didn't help matters much.

In retrospect, trying to see the plus side, perhaps reading *All Quiet on the Western Front* in that environment helped me 'live' the book more than I otherwise would have. Perhaps if I had happened to read it sat in a centrally heated flat back in Aberystwyth I mightn't have empathised with the narrator quite as much. But then, do you really WANT to empathise with an account of that kind? Perhaps. Surely experiences are best when they leave an impression, and thus have 'value'. So, reading this book mid-passage during a stormy season in the North Atlantic made the experience more 'valuable' ... Right? Hmmmm. Discuss.... Certainly, absorbing descriptions of enduring an artillery bombardment whilst sitting in a sopping-wet sleeping bag inside a steel tube being thrashed around in a Force 9 storm atop 1000 metres of angry water most definitely served to heighten my emotional reaction to the story. I really 'felt' the experiences which were so evocatively painted on that book's pages. It was probably a bit like watching *With Nail and I* during an excessively long, hung-over exposure to budget public transport, or reading Shackleton's *South*, or Apsley Cherry Garrard's *The Worst Journey in the World* whilst sitting in a bath of ice cubes with a blob of penguin faeces smeared across your upper lip.... In fact, if you really want the 'scratch and sniff' experience of early twentieth century trench warfare I can think of few better simulations than reading *All Quiet on the Western Front* whilst at sea in the midst of a gale.

Thus passed our first day en route from Nuuk to Oban. There we were: three lads sitting in the lazy sun, reading books and contemplating our own mortality. We were so excited (or perhaps just distracted) that the simple watch system we had arranged didn't kick-in until after dinner, which was one of Matt's pizza specials.

We settled on a three-hour, one-man watch rotation, i.e. we kept watch alone, overseeing the boat single-handedly for three hours before rousing the next man from whatever he might be doing. This way there was always someone at the helm, twenty-four hours a day, to deal with any changes or problems that might arise (i.e. reefing sails when needed, altering the angle of the sails in the face of changing wind direction, generally troubleshooting and, of course, generally taking one for the team and ensuring that at least one of us was always there to make the most of every opportunity to soak up some good, old-fashioned suffering), or simply to look out for other ships which might choose to put themselves in our path.

In reality, you were normally only alone on watch during the night or in bad weather, for on sunny days or in light winds during daylight hours there was usually someone else up and about. This helped when it came to reefing sails or doing work on deck, as it really pays to have someone either mucking in or at least watching your back when you're indulging in a good bit of salt seasoned misery at the mast. Of course, when conditions are particularly bad and the sails needed tending, we all accepted that we should (and would) be called up and out of bed early to lend a hand. That's why we were here after all: to get cold and wet and show the North Atlantic Ocean who was boss! Not to sleep comfortably all the way home. We could have done that on a plane. No?

So, the rota was Sam, myself, and then Matt. Sam would wake me, I would later wake Matt, Matt would subsequently wake Sam, and so on and so on *ad infinitum* or for as long as it took us to get home. We settled on three hour watches. Six hours off in between shifts seemed like time enough to get a useful amount of sleep, and also it ensured that we would be on watch at different times every day. Had we gone with watches of four hours duration each of us would always have been on at the same time of day, that is for example from 12 noon until 4 pm every day and 12 midnight until 4 am every night. Three hour watches added variety. Everyone would get their share of sunrises, sunsets, afternoon sun and inky darkness. Five hour watches would have provided this variety as well, but five hours is a long time to be at the helm alone in a night storm! So, in the style of *Monty Python*, three would be the hours of the watching, and the hours of the watching would be three. Watchest thou not to four hours, nor to two, except when then progressing on to three. Five hours was right out!

We also started keeping time by the universal clock, known as 'UTC' (Coordinated Universal Time) or 'Zulu', which is the essentially the same as Greenwich Mean Time (GMT), just a little less ambiguous. This was three hours ahead of the time in western Greenland, but was in line with the times used by coastguard stations to broadcast weather information on the VHF, as well as the weather charts which we were able to download over the satellite phone. It is this 'UTC' or 'ZULU' time with which all of our log entries are kept. So, as the first day came to a close and the first watchman came on duty Gambo's log received its second entry of the voyage:

15/09/10
0300 UTC. Sam:

Sea calm (pond).... Motor. Clear skies, approaching low stratus clouds to south, waves (swell) from SW. Red sky, cloud in W/SW. NO WIND!! Brilliant aurora (good nav). Geoff says 'No speed' every 10 mins. Keeps pilot though.

The occasional temperamentality on the part of Geoffrey that Sam refers to never really got solved, but was likely down to some occasional communication issue between the autopilot and the GPS. The problem was only ever fleeting, and was usually resolved simply by relieving Geoffrey for a few seconds before putting him back in charge (i.e. the good old 'switch of and switch on again' routine). If anything, it really only made him a little more human, and therefore a fuller and more completely anthropomorphised member of the crew. No-one is perfect, right? Even machines get disoriented sometimes! And those first few calm days really were disorienting. Although none of us were really in a hurry to be tossed around, the persistent lack of wind and the continual, protracted drone of the engine created a certain malaise that I perhaps wasn't fully aware of at the time, but in retrospect my memories of that initial southbound cruising phase are almost fluid, as if I had been slightly drunk the whole time. Some things I do remember vividly though are the truly unforgettable sunrises over the Greenland Ice Sheet to our port side, the atmospheric phenomena we witnessed as this persistent, high pressure weather system passed over, our encounters with local wildlife and the experience of being on watch at night and enjoying the ethereal hues of blue and green fluorescence which Gambo left in her wake as our propeller churned us southward through the cold Labrador currents:

15/09/10
0700 UTC. Colin:

Speed ~ 5.0 kts from 0700 again. Tidal currents? Still placid water with zero wind. Good fluorescence, and good aurora last night around 0000 hrs. Course corrected to 150° @ 0730. Spd <5 kts.

Light wind from SE @ 0900. Staysail up @ 0915. 8 – kts variable. Full sail attempted @ 0940. Speed 5.8 kts. VMG 2.6. Aborted. Motoring.

DTD: 1733. Word from Nolwenn: "No wind!"*

The latter log entry (penned by myself) very representatively reads as coming from the mouth of someone who is feeling somewhat in awe of the vastness of his surroundings, increasingly aware of his own inexperience, and who is unconsciously making damn sure to distract himself by focussing on every passing point of interest presented. It also mentions contact with Nolwenn, who by this time was back in France and was no doubt nervously following our progress remotely. We were updating him daily via the satellite phone's text messaging facility, informing him of our location and progress. Nolwenn would reciprocate by checking forecasts and offering his perspective on our options. He did this throughout the voyage, which gave us a tremendous sense of security in the context. He also brought in a friend of his who worked as a professional weather coach for seafarers.

Around about this time we were also collectively growing increasingly cautious of the facts that A: engines use fuel; B: that fuel was finite; and C: there aren't gas stations in the middle of the North Atlantic. We had plenty left at this stage, but also knew that our reserves were finite.

In order to keep track of how much fuel we were using we ran the engine from a small 'day tank' which could be re-filled via an electric pump that drew from the main stash which sat in a holding tank just above Gambo's keel. This day tank could support approximately twenty-four hours of motoring, depending on the RPMs (engine speed) we chose to operate at, and we knew that there was fuel in the main tank sufficient for only about six refills. Later on this first full day spent under-way we received our inaugural reminder that our payload of fossiliferous go-juice wouldn't last forever.

15/09/10
2000 UTC. Colin:

First glimpse of level in day diesel tank. 5.8 kts, track 154°. Wind 6 kts variable (port quarter).

My unwritten subtext here reads: *Fuel is running out! We need enough to get us out of trouble if and when everything goes tits-up later! Where is the bloody wind!*

11 / Restless at sea

Sailing gives you a strange perspective on the world. You see mapped landmasses from the 'outside'. Viewing land from the ocean somehow inverts the way you perceive coastlines. Peninsulas are no longer walkways into another realm, but nasty protruding obstacles to be avoided. Enclosed bays aren't invasive intrusions of water which force human traffic on long, circuitous overland diversions, but rather sheltered havens and anchorages where a ship can hide from the ravages of open-water weather systems. Everything looks different from the sea.

It wasn't until I began to sail around the United Kingdom from 2011 onwards that I really began to genuinely 'see' the outline of my own country. I felt the same way as we motored down Greenland's coast: really 'seeing' the Davis Straits, and 'feeling' the Labrador currents first-hand for the first time. I felt like I myself, and my boat mates to boot, were very much part of the map.

Furthermore, sailing places you in something of a unique category as far as ocean residents go.... Pretty much everything else which lives in the maritime environment either flies above it or swims below it. As 'beings' that simply float along, ships and boats are almost alone, save for the occasional log, coconut, piece of man-made rubbish or, in the Arctic regions, polar bear, which perhaps surprisingly to some is a truly marine mammal which can happily swim hundreds of kilometres in freezing waters.

On a sailboat you slide along the transitional zone between two universes and only occasionally do you encounter occupants of either one. It can be very lonely, living in the seam between two masses separated only by their relative densities.

Every so often though you have an encounter which reminds you that, despite all appearances, you are never really entirely alone. My first encounter of this kind, after leaving Nuuk, occurred at some point in the evening of our second day. I was alone on watch as both Matt and Sam had retreated below for some shut-eye. We were still running under engine, and Geoffrey was, as ever, doing an admirable job of keeping us on the straight and narrow. All there was for me to do was sit in the 'gin and tonic seat' and keep an eye out for any obstacles or changes in conditions which might force us to actually do something. It was pretty peaceful.

The gin and tonic seat was just a small padded platform built into the chest-height railings on Gambo's stern. When sitting in it you had a great view over the cockpit dodger and across the deck. Your rear end was actually overhanging the boat's transom, and so you could also look sidewards and down onto the water being swept out in our wake. It was a great place to sit, especially as at this time in 2010, unlike back in the days of my Antarctic Gambo initiation, the boat's stern featured a high and robustly constructed A-frame. This A-frame was useful not only for mounting radar receivers, GPS antenna and other assorted gadgets on, but also for simply clinging to in times when the preferably intimate relationship between foot and deck (or indeed between arse and gin throne) seemed in any way threatened.

So, ensconced as I was on the back deck Pimm's perch, peering out across the deck and towards the horizon as the light slowly ebbed from the sky, I was absent-mindedly scouring the ever-approaching seascape for danger, when quite suddenly and unexpectedly a large log slid into view just off our starboard side. It must have been virtually off our bow as

I hadn't noticed it earlier. The boat's mast and forward-mounted genoa furling rig could be something of an impediment to visibility in the 'twelve o'clock position', meaning that, frustratingly, the point where any yacht helmsman's view is poorest lies directly ahead! We must have been virtually on a collision course with this log for some time. Had I been standing on the bow I would likely have spotted it, but from the helmsman's position it had gone unnoticed by me until now. It occurred to me how glad I was to be sailing in a steel boat, and how much more nervous I would be of hitting debris at sea in a fibreglass or plastic boat. A tree trunk would make one almighty bang if we were to hit it in the wrong way on Gambo, and we'd feel the impact for sure, but her metal hull would almost certainly hold. A fibreglass hull on the other hand....

These thoughts were running through my head while I turned to watch this transient tree trunk track along into our wake when suddenly there was a fierce whooshing sound, accompanied by an eruption of water vapour which exploded at a slight angle from one end of the 'log'. As I watched, the explosive end of the log lifted slightly out of the water before dipping down low and out of sight beneath the placid seas. The other end rolled forwards briefly before revealing a relatively small dorsal fin and, finally, after what seemed like a very long second later, a curvaceous tail fluke which was lifted high into the air before slipping delicately into the cold dark waters behind Gambo. I had very, VERY nearly ploughed Gambo clean into the side of a resting, or 'logging' sperm whale!

This was a real thrill, to pass so close to such a large and iconic animal. The sperm whale is just so unlike any other whale, both in its appearance and its habits. Encountering them always brings a 'here be monsters' feeling to the pit of my stomach. Sperm whales are by far the deepest diver of all the cetaceans, plunging to depths of thousands of meters in search of their primary prey: the giant squid. The very idea of this whale hunting at the bottom of the world's oceans, doing battle with colossal squid, equal almost in length to the whale it's-self... It's almost the stuff of fantasy. And of course the sperm whale is no stranger to literary fiction. It was a sperm whale after all – a white sperm whale – which drove Captain Ahab to the brink of madness in Herman Melville's famous tale, *Moby-Dick*.

So, as Gambo cruised on and away from this now hidden beast, adventure was certainly in the air.

15/09/10
2000 UTC. Colin:

Passed 20m from dozing Sperm Whale. Large!

Added 'points of interest' to GPS: Continental shelf; Mid-Atlantic Ridge; Rockall.

Smoke from exhaust. Engine temps: Water: 85°c; Oil pressure: 60 PSI @ 2008 UTC.

Later, when I told Matt and Sam about the 'whale log' encounter I got a bit of a ribbing for not shouting them both up on deck for a look, but in reality you often don't have time to call others out. Often all you see of wildlife on the ocean is a fleeting glimpse, and by the time you rally your troops for a photo shoot the action is long over. True, sometimes things go your way and whales or dolphins can be very inquisitive and sociable. But these moments are quite rare, and often you need to go out of your way either to create them or simply to make the most of them. Our agenda was simple: We were going home! If we stopped moving it would be because something had gone wrong.

So, my little brush with nature behind me, I looked for some other things to keep me occupied before rousing Matt from his berth to take over on watch. As I mention in my last log entry (above) I decided to plot up some 'points of interest' on our GPS which might raise spirits over the coming weeks. After all, we were tourists in the North Atlantic, and every tourist needs a map of memorable landmarks to visit during their time 'abroad'. Of course, open oceans aren't known for classical architecture, art galleries, top-end shopping districts or indeed breath-taking landscapes (although I'd say they actually abound with the latter if you choose to see things that way). But, still, I summoned up imagination enough to add three new points to the itinerary.

The first point I added was the continental shelf. We would soon be crossing this as we made progress further south, eventually leaving Greenland behind and moving forward into the much deeper waters of the open ocean.

Continental shelves are found off the oceanward coast of every major landmass on Earth and are defined as a boundary where the seabed suddenly drops off from being relatively shallow to lying at depths of several hundred or, in some places, even thousands of meters. These continental shelf boundaries mark a transition between rock types. On land, and on the seabed in shallower, coastal areas, the rocks beneath us are generally very old indeed, or at least the rocks we can see on the surface are lying on top of much older rocks which have existed for aeons. Of course the rocks on the surface of our planet are always changing, and always moving around. Even whole continents are in constant motion. They occasionally collide and join together, only to break apart and drift in separate directions many millions of years later. This is called continental drift, a system driven by convection currents within the Earth's molten mantle which, through the movement of hotter and colder bodies of liquid rock, literally drag the surface of our world around, destroying that surface in some places where continental masses (known as 'plates') collide, and creating new land or new rocks in other places where lava and other volcanic emissions, essentially the life blood of the planet, emerge, or erupt, to fill the gaps left behind as these plates move apart. So, as James Hutton (an eighteenth century Scotsman, often referred to as the 'father of modern geology') once postulated: "… in some places our world is continually being destroyed only to be renewed elsewhere"; a controversial view in times when science was obliged to toe the religious line!

So, continental shelves mark the edge of an old continental landmass. The deeper water beyond the shelf sits on newer rocks, created through volcanic activity as two ancient landmasses drifted apart. It is younger material like this which makes up almost all of planet Earth's seabeds. So, crossing the continental shelf seemed worthy of mention on our planned route. We would, from a certain perspective, be travelling through time. Not only that, we would very quickly find ourselves floating not in 100 meters of water, but in over 1000 meters of dark, cold, uninviting, inky death. It would be a psychological event, marking a real threshold in the voyage.

You know once you've crossed the continental shelf that s**t just got real. There are, from that point, no harbours nearby and very little in the way of ready assistance. If you sink, it's a long way down. When the Titanic sank in 1912 (hit by an iceberg most likely of Greenlandic manufacture) she went down in around 3800 meters of water and likely took something approaching ten minutes to reach the seabed. It's a scary thought to dwell on, and although from our perspectives as sailors there really ought not to be any difference between sinking in 40 meters of water and sinking in 4000 meters of water, both being equally likely to ruin your day, the thought of sinking in deeper water is somehow worse. So, yeah, crossing 'the shelf' would be a memorable moment. Furthermore, sea conditions are generally likely to get worse in deeper and more open water: with more open space and no landmasses in the way, winds have much greater 'reach' over the water. On the open ocean, wind has more room to blow, picking up waves and coaxing them up to greater heights over larger distances. For a sailor these are all game-changing factors.

Anyway, beyond the continental shelf I struggled to find anything else of real note on the charts for the best part of 1000 nautical miles along our planned route; that is, until the point at which we would be more or less south-west of Iceland. At this juncture we would cross the Mid-Atlantic Ridge. This was my next ship's track tourism tick-off.

Following on from what I've already said about continental shelves, i.e. that the Earth's continental plates are in constant motion and that the surface, or 'crust' of our planet is being continually and simultaneously destroyed and renewed (or essentially 'recycled') along zones of plate collision and zones of plate divergence respectively, the Mid-Atlantic Ridge is the seabed manifestation of one such zone of plate divergence. Basically, it's a seam on the Earth's crust where plates are moving apart, creating a long crack which runs from the high latitude Arctic waters of the North Atlantic to the Antarctic's Southern Ocean. Into this crack fresh rock (in the form of liquid lava) is constantly being injected from the Earth's mantle below. This crack is perpetually opening as the North American and South

American continental plates move ever further westwards, away from the European and African plates which themselves are in constant motion eastwards. Here, eruptive lava has built a colossally long mountain chain which bisects both the North Atlantic and the South Atlantic, boasting some truly gigantic peaks. Of course these peaks are invisible to sailors, even some of the most prominent amongst them being submerged beneath almost 2000 meters of water, the exception being the country of Iceland which is in fact part of this most lengthy of volcanic mountain ranges.

Anyway, being a bit mountain mad myself, and having lived most of my life in awe of the Atlantic Ocean, the notion of sailing over the top of this colossal but invisible range of sinister, hidden peaks fired my imagination in no uncertain terms. I've always found submerged wrecks and underwater landscapes somehow terrifying, even if only on a low-order or background level, so the idea of being suspended directly over such herculean hills, with over 1000 meters of water between our little steel tube and their remote summits, filled me with genuine awe. Also, crossing this threshold would mark our transition from sailing in North America to sailing in Europe. Beyond this ridge we would be back in our home continent! At least from a geological perspective. So, I plotted the 'most likely' point at which we would cross this hidden barrier and looked forward to the time when we would (hopefully) ride past it.

Finally, I added a point entitled 'Rockall'. Unlike the previous two attractions, this little tourist temptation really was just a point on the chart, and not a ridge or linear barrier which we were certain to cross at some point. So, I knew it was a long shot that we'd pass close to it, but I marked it down nonetheless.

Rockall is a small sea stack composed of hard, granitic rock which sticks abruptly and incongruously out of the Atlantic Ocean's punishing waters some 160 nautical miles west of Scotland's Outer Hebrides. Although its sovereignty has been disputed in the past, this forsaken spike of rock is generally regarded as being the most westerly point of the United Kingdom. It is completely inaccessible by conventional means, which is understandable given its location and that aside from a few sea birds and the consequent layer of dried guano which clings to Rockall's 20-meter high 'summit', there is literally nothing to see! But, still. In the vein of a true Mallory-esque tradition, I naturally felt a very strong impulse to see Rockall with my own two eyes, simply *"because it's there'.* Whether I'd be able to would of course be down to the winds, which we would experience over the coming days and weeks; these winds determining the exact direction from which we would eventually end up approaching British waters.

12 / Into 'deep water'

In the few years which have elapsed since Matt, Sam and I completed our Atlantic odyssey I have recalled pieces and parts of the story to many people, both close friends and interested acquaintances. People's reactions have ranged between mirth, awe, envy and, in a couple of cases, incompletely veiled contempt. The reasoning offered by individuals who erect their tent in this latter camp of thought is that we were incomparably stupid to even have attempted such an undertaking given our collective lack of sailing experience. Although in these cases I will generally at least attempt to salvage my situation by recalling previous experiences of sailing in the Antarctic, my years spent as crew with Britain's Royal National Lifeboat Institution (RNLI); Matt's experience of small craft sailing and remote areas mountaineering; and Sam's record as a four season mountaineer, an assistant explosives technician for British Antarctic Survey, and ropes access specialist for the Greenland ice sheet scenes in the BBC's 'Frozen Planet' project…. I believe that between us we boasted a good breadth of experience. But, when the cards are all down, I can't avoid the fact that ultimately, these people are right. We were hideously inexperienced and all-round unprepared for the task of negotiating a North Atlantic crossing, particularly one in late September coming into October. Since then I have bought my own boat and have sailed a fair amount in her around the British coastline and beyond to the Channel Islands; all relatively technical areas to sail. So, my knowledge of sailing and navigation has increased markedly. Now, when I think back to what I knew in 2010, or more to the point what I *didn't* know in 2010, I can't help but laugh. I think that's a healthy reaction, given that we obviously survived to tell the tale. Some of our log entries in particular really demonstrate how much we still had to learn. In the dying minutes of September the 15th, Matt penned one such entry:

15/09/10
2347 UTC. Matt:

Arctic Pilot Book Vol. III (p21) shows currents on the west coast of Greenland. Comparing this with our instruments, it looks to coincide with our changes in speed. Deeper water has slower currents.... [Apparent wind = engine speed].

The somewhat shameful truth is that over the course of the previous two days we had been constantly aware that our speed seemed to fluctuate at regular intervals, sometimes waxing and sometimes waning. Our engine RPMs had stayed constant as far as we were aware, but still speed varied. We weren't exactly sure what was going on, but it was early days. I had postulated *"tidal currents?"* in an earlier log entry, but even after a couple of nights motoring we still hadn't broached the subject with each other. It just slipped under the radar. Not until some bright spark opened the pilot book for these waters – doubtless striving to fill a few moments of an otherwise uneventful watch on windless seas – and we stumbled across tidal charts for the area, did the penny finally drop. Of course it was the tides! These days I would have them plotted down and would know when to expect current changes and reversals in the tidal stream, especially in coastal waters! A course can, and should, be plotted to take into account changes in the tidal stream. Back then though, it just went unsaid. Of course by this time it was really a moot point, as we were almost in the open waters south of the Davis Strait where tides and the like would make little or no difference to our course or to our mean speed over ground. But still, our understanding of just how much of an impact on shipping the tides can have took an emphatic step forward. Our understanding of 'apparent wind' also 'ripened' somewhat after a couple of failed attempts to make way under sail: we had been motoring along on a couple of occasions, making a good speed of around 7 knots (presumably with the tide behind us!) and could feel a good, solid breeze on our faces. The anemometer boasted a muscular 10 knots of wind which seemed to be coming from the port bow quarter. So, we deployed the sails and cut the engine, looking forward to the tranquillity of making way under the breath of the skies. But, unsurprisingly (in retrospect), as soon as the engine was shut off the breeze mysteriously fizzled and died. 'Apparent wind', it turns out, is the cumulative expression of the actual wind speed in relation to your own motion. As we made 7 knots into a 3 knot breeze, we felt 10 knots. If we had been running at 7 knots WITH a 3 knot breeze we would have felt only 4 knots on the bow. Clever, clever, clever! So, although it felt as we were motoring along that we could be sailing under canvas, we were actually creating that wind artificially.

The worst part (other than the fact that we were all scientists) is that I already had my Day Skipper's licence, and had studied all of these factors in the classroom. But, without a practical examination, and having had a skipper above me to set the rhythm of shipboard operations on every trip I had crewed on to date, I had never needed to tackle these issues off my own back. Now that we were in amongst it on our own we found ourselves turning theories into tried and tested experiences. We were learning fast, and all the while making way further and further south, slowly creeping around Greenland's southern extremity: Cape Farewell.

Morale was high and we were still well rested. Although these first few days of motoring certainly created a level of impatience, and although the constant hum of the engine perhaps augmented the already surreal state of semi-belief we found ourselves in, slowly heading for our now inescapable appointment with the open ocean, we were optimistic in each other's company. Every watch brought something new, and as we moved southwards our environment slowly changed, not least of all for Sam, who had been in Greenland working at and around the research project's base camp up on the ice sheet more or less continuously since late spring, followed by a long stint working and climbing farther north around Uummannaq and Store Glacier, well north of Disko Bay:

16/09/10
0300 UTC (approximately). Sam:

The first time I have seen proper darkness since leaving Aberystwyth in April.

Now, down at a latitude of around 61 degrees north, and moving well into the latter half of September, we enjoyed proper darkness at night, and a good long dose of it too. The sun was setting at around 19:20 and rising again at 06:30. Almost every night we saw the aurora borealis dancing in the skies above us, and I spent a good long while trying (in vain) to capture good pictures of the fluorescence which we left in our wake as the boats propeller agitated the myriad planktonic beasties which fill the rich, aerated waters of the Arctic.

Darkness brings a whole new range of things to see, and a wholly different perspective on your environment. This is something that's easy to take for granted, but when you live for prolonged periods in an environment where darkness is hard (if not impossible) to find, a return to the good old reliable day vs. night routine can be as much of a relief as a hot shower after a long winter motorcycle ride! Sam had been living in orange tents on the ice sheet, which don't block out much light, and then in similar tents and bivvi bags farther north whilst climbing, so had found little respite from the light until coming aboard Gambo only a week or so before. The relief of darkness was beginning to settle on him.

When your daily routine is fairly fixed, and you have little or no freedom to vary your environment, the absence of darkness at night, although spookily beautiful, can put you into something of a trance. You lose some part of your sense of self. Night time gives everyone somewhere to hide, even if we don't think of it that way during our normal lives. The daily setting of the sun gives us all a psychological burrow to retreat to at the end of every day where we can leave the cares of the previous twelve hours behind and indulge other sides of our personalities, be that a walk under the stars, an evening spent on the couch with a loved one, a trip to the pub with friends or simply a night spent in a darkened room with only the flicker of a television to entertain us. Darkness is a blanket. When it's gone I feel like I am somehow exposed, and I operate at a constant background level of tension. So, for Sam, our progress south came as a relief.

Of course, inevitably, our progress farther and farther south brought other changes to our lives as well. Geography is a b**ch that way.

16/09/10
0600 UTC. Colin:

Crossed continental shelf! Now it's deep until Rockall bank!

Constant wind ~8 kts off port quarter. SOG: ~4 – 4.4 kts. Track ~ 150°. Attempted 160° but currents reduced VMG by >0.2 kts to 3.8 kts. Therefore have settled for a more eastward track for time being.

Foggy weather tonight. Could be on account of approach to Cape Desolation. Warmer currents from North Atlantic rounding cape? Must watch radar for icebergs in coming night watches.

So, in the early hours of the 16th of September we crossed that psychological barrier that is the continental shelf and struck out into the dizzy depths of the North Atlantic proper. I made an entry in the log to celebrate.

Colin:

I have hoisted our home-made Greenlandic flag. As we forgot to do it when clearing customs and exiting Nuuk, it seems appropriate now, what with the continental shelf behind us and the full force of the Atlantic threatening. Last radar echoes of the mainland also received this afternoon at over 40 nm distant. Goodbye Greenland!

We were no longer in coastal waters, that was for sure. And it felt like it. Deep water instils a wholly unique brand of vertigo in you. There are few sensations quite like flicking a shiny coin over the side of a boat with a clear kilometre of ocean under the keel and watching it flicker and blink downwards and out of sight. You know that it has many minutes of free-fall ahead of it before it'll finally settle on a patch of loose seafloor sediment which, in all likelihood, will never be seen by human eyes. It gets the heart rate up! The sea can feel like a friend; a companion which holds your boat aloft and lets you ride it home or away, but like a wild animal, if you stop respecting her, she can swallow you up with little warning. From here on in there would be plenty of water to do the swallowing.

Also, from this point on, there would be no shelter from weather systems or waves. When they come you have nowhere to hide. We knew that in the open ocean this would become a factor.

As it happened, the change came right on cue:

16/09/10
1144 UTC. Sam:

Downloaded GRIB (weather forecast) *file. Had to reinstall the modem. Bad Iridium network* (satellite phone) *connection due to incremental com port assignment* (whatever that means)…. *Took 03 hours 28 mins to download.*

Looks like we're going to get some wind….

2221 UTC:

Nolwenn forecasts wind turning to E and then N. 25 – 30 kts in 24 hrs. After that, NE 20 – 25 kts. Suggests we keep S or SSW.

Now, during the early phases of writing this book, three years after we completed the voyage, I visited Sam in Aberystwyth (where he was just finishing off his PhD) to get some of his perspectives and to look at (and copy) some entries from the ships logs which he had rescued from Gambo upon reaching the UK. Flicking through the entries, reading into the tone of each of our scrawlings as the voyage progressed, successive entries in the context of previous ones, and in the context of remembered events.... It was pretty amusing. What was also amusing was the condition of the sheets themselves! We encountered our first 'proper' wind on the 17th of September, and the state of the page that was opened for log entries that day stands testament to how unprepared we were. Most of the pages were moderately dog-eared, but the sheet headed "17/09/10" was in pieces. It was criss-crossed by pronounced folds and the writing in some places was rendered almost illegible by the black marks made by wet boots on the rubber matting which at that time surrounded the chart table by Gambo's companion way. It was like an ancient parchment which had witnessed great and terrible deeds. This all makes sense when one reads the first subsequent log entry of 18/09/10 (Sam writing):

Last night wind picked up gusting 43 kt. We hove-to @ 0330Z. All very tired. Some seasick.... Matt + Colin set the sail so that we sat off the wind + waves.

On the morning of the 16th we downloaded a weather chart (with some difficulty) in the form of a 'GRIB' (GRIdded Binary) file, which we could overlay on an electronic map, showing weather systems for our area. Isobars were displayed, with wind directions indicated by small arrows and wind speeds depicted both by colour and by the number of 'tails' on each directional arrow. 'No wind' shows as white, light winds as light blue, phasing through deep blue into green, yellow, orange and then into nasty reds and, worst case scenario: deep purple. What we saw was a map covering the area from just south-west of Newfoundland, the eastern end of the Labrador coast, north as far as Baffin Island and the southern tip of Greenland. We were more or less bang in the centre of the frame, and just off to our west, hovering slightly east of Newfoundland and covering a pretty intimidating area in each direction, was a fairly unsightly blob of emphatic red denoting wind speeds in excess of 50 knots, or approaching

60 miles per hour. It was heading north east, i.e. directly towards us.

This isn't what we wanted to see.

We tentatively raised a small amount of sail, set the canvas to catch the newly arrived wind which was presently coming at us on our starboard bow, and started making headway, beating into the weather on a close reach (sailing into the wind) in a south-easterly direction.

Initially the wind speed rose only to 15 knots or so; very pleasant sailing weather! But, in line with the GRIB file's predictions, the state of the sky suggested that this was just the beginning. Very soon the numbers on the anemometer display breached 20, and shortly afterwards the agile airs which whipped our rigging were gusting up to and beyond 30 knots.

Wind upon wind upon wind.

Things very swiftly became fairly uncomfortable. We had grown soft, languishing in the flat calm of the Arctic, high pressure system which had accompanied us on our southward jaunt these last few days. A gentler introduction to actual sailing in the North Atlantic would have been appreciated, but not to worry.... The 'bucket of cold water' welcome seemed somehow appropriate.

Conditions aboard rapidly descended into what I can only describe as controlled squalor.

Despite what we thought had been our best efforts to maintain a shipshape living space over the preceding few days, it turns out that when one stows food and equipment in calm seas, where gravity can be relied upon to pull things down towards the floor and not, say, sideways towards the walls, one perhaps isn't making decisions whilst in full possession of the facts.

Gambo's interior became a tip. As darkness fell and we each took our successive turns on watch, water soon began to make its presence felt inside as well as outside of the boat. Every time one of us came down the companionway and into the cabin we brought half a wave with us. Foul coloured liquid, mixed with the myriad bits and pieces which had by this time found their way onto the floor from both the engine compartment and the galley, began to build up on the wooden decking and, what with us being on a starboard tack (which causes any sailboat to lurch continually to port) gravity began pulling everything sideways and brown water was soon swilling in small pools around the corners on the 'downhill' side of the cabin.

Every time one of us came off watch we would deposit our sopping wet foulies in an accessible pile aft of the galley before making our way as rapidly as possible to our bed, ideally before the urge to vomit grew too overwhelming. This habit meant it would be less effort to slip (or, more accurately, squelch) back into our sailing kit on the next watch, but it also ensured that water didn't have to go far to slink from foulies to floor. It also created a bit of an obstacle course between the main cabin (where our berths were) and the companionway, creating optimal conditions for opportunistic aquaplaning, impromptu nose diving, accidental flailing and a fair bit of general shit spreading.

Some of us spent longer periods outside, so thus brought more water in with us when we took shelter to write a log entry or the like. Others spent five minutes up top, checking for other ships or potential trouble, and five minutes below to warm up, this regular alternation again introducing ample occasion for the sea to get inside and visit our odd little floating penal colony. Of course, the regular manufacture of tea, hot chocolate and warm fruit juice (from powder), as well as the sometimes disastrous transfer of said luxury beverages from the galley to the cockpit, also introduced other exciting ingredients to the film of watery grime which soon coated pretty much everything aft of the on-board toilet, which we just won't mention. Both we and the boat were most certainly being 'broken in'.

By 03:00 we were exhausted. The wind was consistently up beyond 30 knots and occasionally gusting above 40. The waves were getting taller and taller, and life had become about as much fun as Stalinism. All three of us were beginning to feel like we'd put to sea in a cocktail shaker rather than in a yacht. So, deciding that perhaps discretion may indeed be the better part of valour, we decided to take the only 'out' available to us, and 'hove-to' as quickly as we could.

Heaving-to, as we learned from consulting *Heavy Weather Sailing* in Gambo's library, involves setting the sails to such an angle that the boat holds both its position in the water and its angle to the wind, most importantly preventing the boat from turning broadside to the waves and thus reducing the risk of capsizing.

Matt and I donned foulies and headed up top, first pointing the boat directly into the weather so as to stall the sails and arrest the forward motion of the boat. The wind whipped around us in the darkness, airborne sea spray coating us in a pervasive film of salty wetness.

We rapidly dropped the mainsail altogether, tying it up tightly so as to prevent excessive flapping which, over prolonged periods, inevitably wears the canvas out. This is a bit of a rule in sailing; if it flaps or rubs it will eventually break, probably sooner rather than later, and almost certainly when you are least prepared for it.

After dropping the mainsail we tweaked the genoa, deploying only a couple of meters, and 'backing' it to the wind, that is to say that we sheeted it, or pulled it tight on the 'wrong' side of the deck, that being the side from which the wind was to come. We were now abeam to the weather, with the genoa full enough to act as a stabiliser, but sheeted so as not to provide any useful amount of forward momentum. And so, we came to a halt, angled more-or-less 45 degrees to the weather, and spent the remainder of the night below deck, in our beds, waiting for the weather system to pass.

Before crawling into my scratcher I got on the radio and sent out a general maritime information message called a 'securitè'. This message – beginning with the words "securitè, securitè, securitè. This is yacht Gambo, yacht Gambo, yacht Gambo on channel 16...." – serves as a general warning to any other ships in the area. In this case it was a navigational warning, for as we were stationary, and our intention was to spend *most* of the next few hours in our beds resting, we constituted a navigational hazard. I relayed our position and stated that we were hove-to in high winds, and therefore we would appreciate it if any ships transiting through our general area would at least *consider* not ploughing us down and sending us under their hull and to the bottom of the Atlantic. It was now my watch, and so I made sure to pop my head out of the hatch every twenty minutes or so to check for other ships, but at this stage our priority was rest.

As soon as we hove-to life became infinitely more bearable. Thanks to the fact that we were no longer beating into the weather the constant concussion of waves on the bow ceased – as did the perpetual lurch to port. This allowed gravity to return to a more traditional floor-ward tug.

We lolled up and down amongst the white horses which were still stampeding past us outside. The 'backed' sails provided us with stability, if not with drive, preventing the boat from rolling from side to side on the incessant white-capped waves which were now taking us at a slight angle off the bow.

We all actually slept very deeply, and as the morning's light began to creep through the narrow windows which encircled the cabin, overwhelming the somewhat seedy, red, chart table light which otherwise provided the main light source during hours of darkness, we felt the strength of the wind ebb away. It's remarkable the effect a good, untroubled sleep can have on the mind; as the sun crept up we all felt compelled to get under-way again as soon as possible. So, grateful for having spared ourselves a night of unnecessary misery, we loosed the mainsail boom, lifted the mainsail onto the third reef, hauled the boom in again to catch the wind and correctly sheeted the genoa, un-stalling the boat and once again creating useful drive, propelling us through the waves anew:

18/09/10
2225 UTC. Sam:

We are now 2225Z sailing east chasing a low pressure system. The sail looks good and we have good VMG (Velocity 'Made Good', i.e. speed in the right direction). *Overcast and raining slightly. Wind 25kt gusting 30kt 030°. We're sailing on a close reach with jib (1m), full stay sail and 3rd reef on the main. BATT 12v 7A discharge 54% 25Ah remaining. 40 hours till discharge.*

13 / Through the pain barrier

The next couple of days passed relatively uneventfully. Other than a momentary 'sense of humour failure' on the part of our fourth and arguably hardest working crewman – Geoffrey – things were settling down nicely. Geoffrey had decided for one reason or another to change the terms on which he would be working with the GPS, and suddenly announced via a series of loud beeps that he was steering, without guidance. He'd stopped listening to the GPS-keyed compass; likely an altercation over working conditions, what with Geoffrey being stuck perpetually above deck. However, after a short bout of involved techno-diplomacy, Sam eventually managed to calm things down and reconciled Geoff to working with the ship's magnetic compass instead. This was actually still keyed into the GPS system, but Geoffrey didn't need to know that. We did of course have an old-school boat compass, but that was for visual reference only, when steering manually, and not something that could in any way be made to communicate with the boat's on-board electronic suite. The generation gap was simply too wide. So, we were soon under-way once more, feeling thankful that a sizeable part of Sam's fieldwork had involved programming GPS sensors on the Greenland Ice Sheet! If it had been only Matt and myself aboard I firmly believe that from that point on we would have been steering manually all the way back to Oban, having made Geoffrey walk the metaphorical plank.

Things seemed positive. We had broken through the pain barrier at some point during sunrise on the 18th of September, and with the moderate baptism of fire which being spat out of the Labrador Current had brought now behind us, we settled into more of a truly seafaring routine:

20/09/10
Time not marked. Matt:

Wind ~25 knts. Speed ~5.5 kts. Heading 085°. Wind from 60 degrees to the bow (020°) NE.

Very smooth. Getting good rest. Might get more sail up tomorrow. Now, Medium jib, all of the staysail, 3rd reef on the main.

We had passed to the east of Cape Farewell, marking another major psychological milestone. Greenland was now off the chart, so to speak, and beyond lay the open ocean. We squared away the cabin in a more nautical fashion, stowing absolutely everything which wasn't actively being used, and soon enough the bilges were brought under control, making the boat's interior a far drier place to be, at least intermittently.

We even found the time and stomach to get back to a bit of reading. I was getting close to finishing *All Quiet on the Western Front*, thank the stars! Nothing quite compares to reading about artillery bombardments, trench conditions and the spectacular variety of ways in which the human body can be dismembered whilst sitting out a severe gale in over 1000 meters depth of water, waves smashing into the hull like irate black rhinos and brown, foul smelling water sloshing around the cabin of one's supposedly watertight sailing boat. No, reading had certainly not been a calming influence during that time.

Meanwhile, Sam had been absorbing *The French Lieutenant's Woman* by John Fowles which, when quizzed on its plot and general content, he described simply as being a "book on Victorian ethics." I'm STILL intrigued!

Matt had been reading Charles Webb's *The Graduate*, which I'll be honest struck me as having the potential to be almost as unsettling on a small yacht in the throes of the North Atlantic as my own suspect choice in 'light literature'. Dustin Hoffman, 1970s America.... The hairstyles and the seduction, intertwined with seasickness, howling gales, pervasive salt water and the ever-present miasma of bilge water and masculine body odour.... Surreal.

Passing through our first rough weather also had the interesting effect of encouraging us all to send messages back home. Short messages albeit, what with our only facility being the satellite phone text message service, but messages which I'm sure were appreciated back in the motherland nonetheless. None of us had sent word homeward thus far. It's hard to explain, but it seemed to me that perhaps writing to loved ones with a positive progress report after only a few days of windless motoring would have been tempting fate. We needed to be tested slightly before it became acceptable to even consider transmitting a textual 'thumbs up' back east. But, with the winds now all around us and the first uncomfortable night behind us, our mothers and partners got their first bolt from the blue. I don't recall exactly what I sent my mother, but I know it included our GPS position. Little did I know at the time, but she plotted every co-ordinate I sent her up on her atlas, and even went as far as to phone the UK Coastguard Agency for some words of encouragement about our chances out there on the gigantic Atlantic.

So, we began to settle into life on the waves. The breeze no longer put the wind up us to the same extent, reefing the sails and working about the deck in rough spots became more of a thrill than an exercise in terror management and we began to really know our way around the boat's systems and idiosyncrasies. I think Matt's log entry of the 20th of September gives a feel for how our relationship with Gambo began to change:

20/09/10
Time not marked. Matt:

The night of the 18th/19th was also rough. Winds up to 44.6 kts around 2 am. Kept sail up to make ground east. It worked. Had too much up at first, but OK once we dropped the staysail, then on 3rd reef on the main and small jib.

1: We've not had the outhaul tight enough on the main reef, so have had intermittent slap on leech.

2: Staysail winch is going to break someone's face (starboard).

3: Wind generator is not enjoying 25 – 30 kt winds. Should be less wind tomorrow so might be able to keep it on.

4: Maybe top up fluid in steering hydraulics tomorrow (Geoffrey), which are getting warm.

Pos: 56° 53.5855' N, 41° 30.3635 W. 02:55Z.

20/09/10
1200 UTC. Matt:

DTD 1152.

Wind ~ 20kts has been 15 – 20. Should die down as high pressure moves away in front leaving low winds this afternoon.

Full main sail, full stay sail, 2/3 jib. 2-3m waves.

There are a few comments in these log entries which are worthy of note. First of all, we had begun to really appreciate the relentless stresses which open-ocean sailing places on the everyday hardware which we wholly relied upon to propel us through the water and ultimately get us home. You need the wind to sail, and you need the sails to catch the wind, but unlike most vehicles, which work hard for relatively short intervals whilst we drive from one place to another, a sailing vessel's essential components work twenty-four hours a day and seven days a week. They are always under stress, therefore it's impossible to over-emphasise how important it is to stay on top of the position, tension and general welfare of every rope, canvas, shackle, pulley and stay. Matt's description of the leech line on the mainsail (the thin cord which runs along the trailing end or 'luff' of the sail) shows that we were beginning to pick-up on the details. This cord was too loose, causing both it and the luff of the sail to flap as air spilled from the back of the canvas. This meant both that the sail wasn't performing as efficiently as it could (bear in mind we still had well over a thousand nautical miles to cover and every knot we could squeeze out of the boat counted towards our eventual return home) and also that it was being un-necessarily battered by the elements. A flapping sail is a sail in pain! So, we duly tightened the leech line, arrested the flap, and from that point onwards knew what to look for, at least on that particular piece of deceptively innocuous looking 'string'.

Some other pieces of hardware which were indispensable, not so much to the performance of the boat but rather to our ability to manage her, were the winches. Gambo had six winches. Two large ones were fixed down aft of the companionway dodger (one on either side of the cockpit) and these were used for cranking in the mainsail sheeting lines. One smaller one was located on the starboard side of the cockpit and this chap's job was to assist us in furling and unfurling the genoa. Two were also situated on the sides of the mast; one on port and one on starboard; and these were used for hoisting the mainsail and the genoa and/or the staysail respectively. Finally, we had a smaller winch at the base of the mast, facing aft, which was used for tightening the mainsail reefing lines. These winches were of variable vintage, but the newest of them had been fitted in 2004 in Piriapolis, Uruguay. That was six years prior to our current adventure, so it's fair to say that each winch had seen its fair share of action. We had serviced them all before leaving Nuuk – dismantling them, cleaning them, rubbing down corroded surfaces and administering a liberal quantity of marine grease –

but when things get to a certain age no amount of grease will make them new again.

As Matt mentions, one winch in particular was beginning to develop a little too much personality in its old age, namely the winch on the starboard side of the mast which tended to the staysail and genoa. Generally speaking, drum winches function either on a one-way ratchet (which allows you to crank a rope under tension but prevents it slipping backwards) or on a two-way double 'gear' system which cranks the rope in one direction but at different speeds depending on whether you turn the winch handle clockwise or anti-clockwise. Both are great labour-saving systems, especially on a boat with a sail area like Gambo's, and both have the safety feature of spinning in one direction only. Well, they're SUPPOSED to spin in one direction only. The staysail winch took exception to this convention and decided it wanted to swing both ways, so to speak. Its ratchet system was shot, meaning that although it still worked, and could be cranked by hand, you needed to keep your hands on the winch handle and keep it under tension AT ALL TIMES, lest it spin back on you. A tensioned rope, attached to a large sail filled with the breath of the Atlantic, is a veritable powder keg of potential energy, and when Matt says that the *"staysail winch is going to break someone's face…."* he isn't exaggerating. We always operated that winch as a pair; one set of hands to turn and control the handle and another set of hands to 'tail' or control the loose rope downstream of the winch drum, keeping it under tension so as to allow the handle cranker a little margin for error, and also to relieve him of the virtually impossible task of safely cleating the rope without releasing too much tension on the winch handle. If this went wrong, it was very easy to imagine suffering a fractured skull from a winch handle spinning back into your forehead at something in the order of 40 miles per hour. This isn't so much of a drama in calm weather, but try for a moment to imagine tackling this theoretically simple task in complete darkness, illuminated only by a head torch, bracing yourself against 50 knots of wind, riding through waves with heights in excess of 7 meters, with wet hands, a wet winch handle and gravity which is about as predictable as the DOW Jones index…. I still shudder just thinking about that winch.

The next piece of equipment which Matt mentions is the wind generator.

This bit of kit took the form of a moderately proportioned windmill which was mounted on a tall stainless steel pole on the port side of Gambo's stern A-frame. In light winds this windmill generated a good amount of power which kept our on-board batteries charged up, allowing us use of important pieces of electronic equipment such as the VHF radio, the GPS, the radar, the boat's night navigation lights, the cabin lighting, our satellite phone and the small laptop which we used to view navigational information and weather charts, which as you will have gathered we downloaded via the satellite phone. Furthermore, without the batteries we wouldn't be able to start the engine which had an electric ignition. All in all, electricity was fairly indispensable on our twenty-first century venture!

The wind turbine itself had been purchased and fitted in South America the previous year, the whole process having been overseen by Nolwenn, the absent skipper, who would always be proud to tell you how well suited this particular model was for Gambo, and how perfectly balanced the blades were to achieve optimal voltage with minimal component fatigue. And, he was right! It was a fantastic device which looked after us very well. The only problem it had was that in strong winds the turbine's internal coils which generate power could and would overheat, meaning that it could only really be used in lighter winds. As we hadn't seen many light winds so far; only completely absent winds and subsequently fairly stormy winds; the turbine hadn't had a chance to really shine. Fortunately, it came fitted with a fantastic little feature in the form of a magnetic brake which allowed us to immobilise the spinning blades at the touch of a button, preventing it from being damaged in storms. This innovative, extra braking feature might seem fairly sensible, and perhaps something which every wind turbine should have in some shape or form, but Gambo's turbine hadn't always had one. In its previous incarnation, the wind turbine had been fixed to a far less elevated arrangement of stainless steel railings. This placed its often frenetically active array of blades only slightly above average crew head height which, with blades being more or less invisible at speeds consistent with your garden-variety, open-ocean winds, made moving around the stern of the yacht quite the adventure. Moreover, this older model didn't feature the user-friendly magnetic brake. So, in particularly fierce conditions, when the turbine needed to be immobilised and prevented from liberating itself in a violent fashion, we would routinely tackle it with a lasso and a fair dose of trepidation, literally tying the monster up against the railings.

Anyway, we were becoming slightly concerned that we hadn't had much opportunity to properly charge the batteries since the engine (which also charged the boat's systems) had been switched off. This meant that we had to be particularly frugal with on-board lighting, only use the radar in particularly bad visibility, day or night, and be very careful about how high the brightness on the GPS display was set. Little things really, but every volt counts when you live in such a self-contained environment. Hopefully the winds would drop soon, allowing us more rest, more sail area, a touch less discomfort (as the boat would heel over less through weather-helm) and a bit of battery juicing.

Matt's final comment about the hydraulics, via which Geoffrey steered the boat, seems like little more than a side note, and we duly topped up the fluid, but this little niggle would come back to haunt us later in the voyage.

14 / Wear and tear

So, in the fullness of time, weather comes and weather goes. Later on the 20th of September the winds lightened, permitting a few hours of useful wind-genny power capture. But the wind, like the ocean, is a fickle mistress, and after only a relatively short window of modest blow the breeze tailed away to almost nothing and we found ourselves wallowing in the swell with the sails flapping impotently from side to side. One option would be to drop the canvas and simply wait for the winds to return, like the sailing ships of old would have been obliged to do, back in the days when Atlantic crossings could take even large ships months to complete. But this was the twenty-first century! So, we stoked the furnaces and awoke the diesel guzzling go-box for a spot of motoring pending the arrival of the next low pressure system which would furnish us once more with moving airs.

Nothing comes for free. Sails can rip and ropes can chafe through, making wind propulsion a constant exercise in damage control, but engines aren't infallible either. Moving parts, high temperatures and long running hours inevitably lead to problems over a long enough time frame, and before long we found cause for concern with the engine:

20/09/10
Time illegible. Colin:

Sam detects suspicious smell similar to burning rubber coming from engine. All stopped. Engine excavated and alternator belts replaced. Small pooling of oil/fuel also detected around fore (1st) injector. Cleaned, and fuel line tightened. Operation lasted <1hr. Opportunity taken to conduct thorough clean-up of the living quarters decking and the hatch seals were re-tightened.

First engine drama! 230 nautical miles from land! However, now only 188 NM to Europe (i.e. the Mid-Atlantic Ridge)*!!!"*

So, nothing too serious, but another reminder that our ability to get home depended entirely on caring for the boat's systems. We did NOT want to have to rely wholly on wind, or to arrive back in Scotland at some unseen juncture in the future and have to navigate the narrow tidal waterways of Scotland's west coast, far less the busy, traffic jammed locale of Oban Marina, without diesel power on our side. A seasoned sailor would doubtless have fared admirably amongst the buoys and boats of Oban Bay with nothing but a tattered sail and a sheeting line between his teeth, but I knew without a wisp of a doubt that *we* would be in trouble. Somewhat amusingly, in retrospect, Matt's next entry reads:

21/09/10
0550 UTC. Matt:

Fuel is going down, FAST!!! Where is the air intake?! Check rocker set screw tightness, loose after adjustment? Burn rich? Leak on Exhaust? Check filters; transmission box oil level???"

He (we) may have been a little nervous.

This period of windlessness lasted longer than we expected, resulting in a full night of relatively anxiety-free motor cruising on stable seas and with no sails to tend. By the next morning however a breeze was beginning to pick up and we were able to hoist the sheets once again and take stock of our situation:

21/09/10
0708 UTC. Sam:

Wind is currently 15kts from 100°. Engine has been running 15h 16 mins since starting or 81 hrs 26 mins since refuelling in Nuuk. We have 7x 24 hrs of fuel plus emergency Jerry cans – so we have used just under half.

Batt #2: 14V 4.9A charge 216AH charge 101% Fully charged.

Batt #1: Going to charge Batt #1 as a precaution for 1 hour.

Batt #2 Discharging @ 5A; 108 hours to discharge – will switch wind genny on for a bit. Now Batt #2 discharging @ 0.2A. Engine temperatures and pressures are in the green.

Batt selector to #2.

It's plain that at this stage, once again, we were getting nervous about our fuel reserves, and with us tuning in more and more closely to the variability of the wind we developed an almost obsessive fixation with the state of the batteries. Diesel was becoming limited. We had used almost half of our reserves and still had more than 800 nautical miles to go. Although we were almost level with our next tourist stop, the Mid-Atlantic Ridge (which would mark our arrival in Europe) there was still a long way ahead, and some of the biggest challenges (i.e. coastal sailing in British waters) still awaited us. We needed to conserve power and fuel. Sailing on the open ocean can be terrifyingly exposed, and weather leaves you no place to hide, but on the plus side there is virtually nothing to crash into when the nearest land is hundreds of miles away! Tides become a virtual irrelevance and even other shipping only appears now and then. In fact, we didn't see a single other vessel between crossing the continental shelf off Greenland and eventually reaching the shallows west of the UK! Coastal sailing, by comparison, is rife with challenges: shallow water; semi-submerged rocks; fierce tidal currents; concentrated marine traffic; regulated shipping ways; daytime buoys; night-time light signals…. There is so much to think about! But this was all ahead of us. For now, we just needed to get there. So, power regulation was high on our list of things to obsess about. And let's be honest, when you live on a small boat with only two other people for prolonged periods with no opportunity for respite, you need to obsess about something.

This leads me to think of one other vital on-board system which I haven't yet mentioned but which was crucial to the maintenance of crew welfare. The stereo! Music! Without music all you have to listen to are the sounds of the wind or the engine. The wind either whispers or screams, pouring worry into your mind. *"I'm building!"* she'll shout, or *"I'm dying away!"* she'll lament. Either way, she'll lead you on a merry dance of insomnia, cultivating the perpetual expectation that very soon, at any time, she'll give you cause to don soaking wet clothes once again and get out on-deck to wrestle the sails into a new configuration, appeasing her indecisive but intoxicating temperament. Like a paranoid lover she'll fill your every waking minute with her neediness, demanding your attention at all times. Even when the wind is light and constant your ears fill with the sound of stretching ropes, knocking shackles and flapping canvas…. All the sounds of equipment fatiguing and moving inexorably towards eventual catastrophic failure. The engine is no better, with even the slightest change in tone, pitch or rhythm potentially signalling the impending moment when this humming box of energy will eventually just stop playing your game. I might sound a little over-dramatic about all of this, but when sailing, sounds are everything. So, music can be as refreshing as a cold beer after a long day in the office.

Peaced-out trance music was a favourite. Leftfield; Orbital; Ministry of Sound, Chillout compilations. All good stuff. I also enjoyed a bit of Iron Maiden from time to time. I'm sure that if the stereo were to fail, or power were to run out and cripple the sound system my crewmates wouldn't have missed the Iron Maiden, but then I wouldn't have missed Matt's somewhat unusual obsession with 1960s rock and Johnny Cash. Not to diss Johnny Cash mind you! I love a bit of Cash as much as the next man! However, in the same way as reading about trench warfare in a storm did nothing for my inner calm, the generally fatalistic tone of JC's typical number wasn't really flicking my switches in the face of several hundred nautical miles of intermittent repetitiveness and suffering.

Anyway, the point I was intending to make several paragraphs ago was that the worst-case scenario for power generation was a prolonged period of high winds which would, for sure, furnish us with full sails, but would also render the wind-genny useless. Much to our displeasure, in the early morning of the 21st of September, our eighth day since leaving Nuuk, Sam checked an incoming satellite phone message from Alun, the owner, who was looking at weather forecasts back in the UK. Sam transcribed the text of this message in the log:

21/09/10
0810 UTC. Sam:

Message from Alun on the 21/09/10
Follow wind to E and then NE as wind upping to 35 kt – heave to if necessary but try to escape as far as possible E to 56°N 36°W for 22/09 0000Z (wp escape).

Hurricane preparation 2morrow: prepare boat for 50kt wind SE as this arrives at 22/09 1200Z. Check everything – hatch top, attach everything, take care and enjoy it.

We particularly appreciated the little postscript *"...enjoy it."* It gave us momentary cause to forget about Alun's use of the word 'hurricane'. However, as if he was somehow tuned into our thoughts and thought it great sport to fuel our fears, a second message arrived only moments later:

Maybe not use autopilot during hurricane. Attach wind vane.

Awesome.

15 / Staring down the barrel of Neptune's bellows

It's funny, but although we certainly had the ability to download and view our own weather charts, I have no memory of doing so after reading Alun's message. I'm sure we did, in fact a photograph exists of me looking worriedly at a weather chart, and Sam tells me that this is the self-same forecast, but I just can't remember it myself. Sam refers to a forecast in a log entry below, stating that only 35 knot winds were predicted, but I simply don't recall. Perhaps the picture painted was just too horrible, and my mind has locked the memory off in the 'best not accessed' folder where I can never revisit it. If that's the case though, then I'd like words with whomever is in charge of my mental records department, as there are a number of other infinitely worse things than weather forecasts that I'd like to archive! Still.... No matter. It seems the outlook at the time was clear: we were going to take a bit of a beating. Again!

Now, having already passed through a sizeable 'red blob' on the GRIB chart a few days previously, and having received no such luridly vivid warning from our chums back home on that occasion, we had no alternative but to assume that this storm was going to be one to tell our grandchildren about.

This set some fairly loud alarm bells ringing. It was certainly a Star Trek "Red Alert" situation inside my head!

Our thoughts turned towards how we were going to survive the next few days. We needed to make sure that everything we needed was deployed and accessible. One of the most important things of course, and perhaps one that most of us wouldn't think of straight away, was our steering mechanism. We needed to be able to control the boat in bad weather. Well, we needed control in ANY weather for that matter, but never more so than when the wind was threatening to do some real damage. I think it speaks for Sam's cool attitude and his ability to keep his head under fire that his next log entry after transcribing Alun's warning was to jot down a quick 'what to do in the event of autopilot failure' guide:

21/09/10
1327 UTC. Sam:

Autopilot fixing options:

1) Check 2nd RS232 if heading out. Band 19200; port 1 on blue box. If 'yes' check cable going to autopilot (AP). If 'no' check GPS antenna.

2) If red light out missing GPS connections. On back of blue box try to push cable back in without removing the plug.

3) Should be 2 orange LEDs under GPS labelled A+S. Green or blue light under heading.

4) If not possible to repair connect magnetic compass.

All I can say is that I am forever glad that Sam didn't go overboard and that I never needed to depend on these instructions. I've always considered myself something of a nerd, and not without a certain amount of pride either: the kind of kid who spent his high school lunch breaks in the English Dept. reading cupboard with the other social outcasts playing Advanced

Dungeons and Dragons or the like. That was me! Despite this pedigree however, I evidently just was not, and am not to this day, as fluent in written 'Geek' as Sam. The prospect of helming manually, standing outside in the clutches of a gale, is considerably less frightening to me than whatever it is Sam is trying to say in the above log entry.

Fortunately, for the time being anyway, Geoffrey was working away quite happily. This meant that all three of us could set about tending to other matters. Namely, tidying the place up ready for Armageddon.

The next log entry (again penned by the cool-headed Doyle) describes some of the measures we took:

22/09/10
0005 UTC. Sam:

With a hurricane forecast, or at least the back of one, we prepped the boat for a storm. The parachute anchor was made ready and stowed in a sail bag on the bow. It consists of 70 m of 1" polypropylene rope attached to a heavy duty canvas parachute and 30 m of rope to act as a running pennant to the stem. When this is adjusted right we should heave to at 50° bow to the waves. It's a second line of defence to setting the sails to the heave-to position. Next we battened down the hatches and screwed shut all the lockers in case we roll upside down. All done to loud music of course. We are only forecast 35 kts of wind, and we have sailed in stronger a few days ago, but it is good to be prepared.

The hurricane is forecast to the south of Newfoundland, Canada today, and tracking up to the SW coast of Greenland tomorrow. Fortunately we are quite far east and may miss the 50kt winds.

I love the different facets of Matt's, Sam's and my own personality that come through in all of our log entries. Each entry is like a celebration of how different we each are from one another. This is a strength in an expedition team. Ask any experienced expeditionary leader and they will agree. Having a mixture of temperaments and attitudes brings variety and keeps everyone sane. Having a number of perspectives prevents any individual from getting too low, or indeed too high. It helps keep things in proportion. Sam is so meticulous, and so understated, both in what he does and how he phrases things. I have a tendency to be perhaps a little emotional, and to over-analyse things both in the heat of the moment and in retrospect. Matt is bubbly and wonderfully positive, but maybe a little impulsive by his own admission. Sam on the other hand can be almost overly straight-forward about things. For example, he easily breezes over the fact that we screwed down all the compartments, just *"in case we roll upside down."* In case we roll upside down!!! Surely this needs some emphasis, Sam.

Actually, 'rolling' isn't as ridiculous a prospect as it seems, for as we already knew, Gambo had a track record on this front. Alun had often regaled us with the tale of when, in the relatively early days of his union with Gambo, back in the nineties, he and his crew of friends and acquaintances had hit some bad weather in the South Pacific during a passage between New Zealand and South America. As Alun tells it, Gambo was hit by a fierce gust of wind and a 'rogue wave' simultaneously, these two entities collaborating to roll Gambo over, almost completely inverting her. Fortunately she righted herself after only a moment or two upside-down and as no-one had been on deck at the time there was no loss of life. Apparently only bruises, chaos below deck and universally shredded nerves remained to tell of what had just happened.

Back then Gambo had been a little different: she'd sported two masts, boasting an aft 'mizzen' in addition to the central main. She'd been what's known as a 'ketch' in this previous incarnation. Now, in 2010, the mizzen had been removed, transforming Gambo into a sleeker and simpler single-masted 'sloop'. So, maybe things would be different! We didn't *think* we would roll, or at least we didn't want to admit to ourselves that we thought we might…. Regardless, there's nothing quite like a bad precedent to fuel the imagination.

a) A rebel in the making: The 125cc motorcycle on which many poorly conceived ideas were formulated. This picture shows her in a fairly representative state of dis-repair outside my flat in Aberystwyth, Wales.

b) The view eastwards from the hills near Kangerlussuaq, looking out over braided meltwater rivers and the edge of Greenland's vast inland ice sheet.

c) Kangerlussuaq 'high street'. These prefabricated blocks house the heart of Kanger's infrastructure. Photo taken outside Kangerlussuaq International Science Support (KISS).

d) Nuuk, looking eastwards towards the more inland mountains. Gambo's mast can be seen on the left.

e) Gambo tied up alongside a large fishing boat in Nuuk harbour.

f) Nolwenn (Gambo's skipper) (left) and Matt (right) chatting over Gambo's chart table after lunch. Nolwenn skippered us as far as Nuuk but would soon leave for France.

g) Before leaving us to our fate Nolwenn did everything he could to prepare us for what lay ahead. In this picture he is servicing Gambo's engine, going as far as to adjust the tappet spacings and valve clearances. You'll notice that he's doing all of this in his underwear, apparently to save his trousers from getting oily.

h) Eventually the time came to leave, and Sam, Matt and I took our leap of faith. This picture shows me (right) helming Gambo out of Nuuk harbour as Sam (left) brings in the fenders. Photo by Matt Burdekin.

i) Samuel Doyle (left) and Matthew Burdekin (right), leaving Nuuk.

j) Sam cooking up a storm in the galley area. Note the angle of the stove, showing where the horizon really lies…. The picture looks towards the stern, showing the GPS display and the companionway in the background.

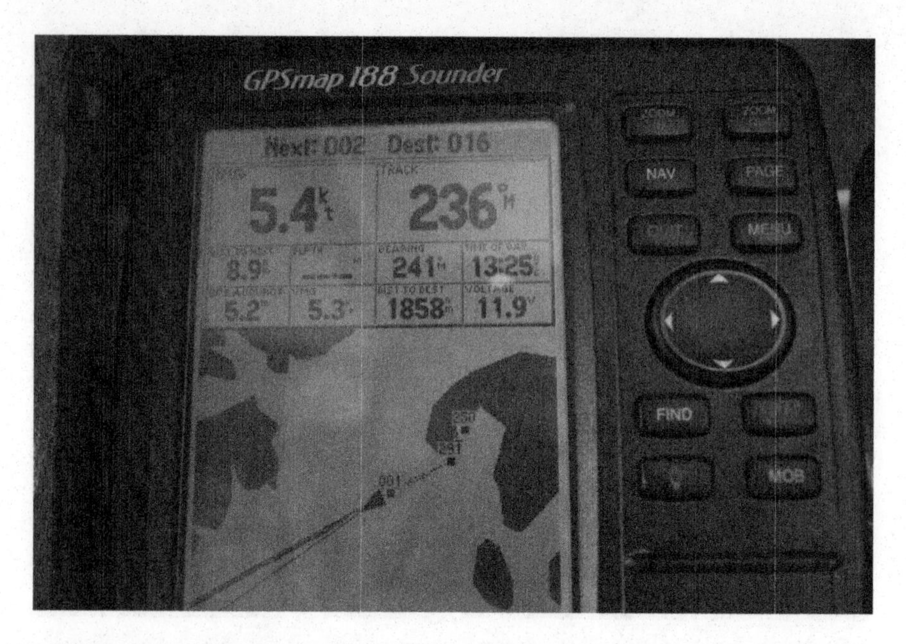

k) Our GPS display after I plotted in a handful of waypoints, including our final destination, Oban. At this stage, just outside Nuuk, Oban is 1858 nautical miles distant, equivalent to 2138 statute miles or 3441 km.

l) Gambo under sail.

m) Matt is a very photogenic chap, showing us in this picture how much fun sailing can sometimes be.

n) At other times however, it's far less fun.

o) Every time the weather changes, the sails need to be altered. This can be very uncomfortable and occasionally quite dangerous. This picture shows me reefing the mainsail in a moderate breeze. Photo by Sam.

p) Me, looking a little worse for wear after a 'spicy' trip up on deck to reef the sails. Photo by Sam Doyle.

q) Sam gets his first ever scuba diving lesson after Matt's 'refresher session' beneath Gambo's hull. Photo by Matt Burdekin.

r) Taking a quick dip in the middle of the North Atlantic Ocean.

s) Matt's famous Atlantic pizza: the king of all morale boosters.

t) Sam has a 'Men's Health' moment and decides to have a salt water wash a few days short of Oban. Note the tattered remains of our old genoa still lying on deck.

u) Tired, battered and bedraggled, but safe alongside at Oban Marina, the isle of Kerrera. The higher houses in Oban can be seen across the bay in the background. We had made it to Argyll, Scotland; where I was born.

v) The three of us take a moment to enjoy the pleasures of terra firma. For Matt (left) this means mobile phone signal. For Sam (right) this means intense land sickness. Life can be so cruel!

We were already terrified of the ocean, and Alun's text warning had very definitely poured water into the deep-fat-fryer of our fears, so we were taking no chances. What Sam doesn't mention is WHY we were bolting down the hatches. It was because, amongst the usual cargo of tinned food and bric-a-brac, we also had a wealth of tools, several tins of paint and two spare gearboxes in those lockers. None of those things are light in weight, and as if the thought of being upside down in a storm isn't bad enough, imagine doing that with two automobile gearboxes spinning happily around in the air beside your head! Not a pretty thought. So, we figured it best to put them under house-arrest. Even tinned food wouldn't make much of a travelling companion in washing machine conditions for that matter! So, we gladly made the decision to forego eating canned produce for a few days as the screws went into the locker lids. I really don't know if those screws would have done us any good, but if they did nothing else they at least made us feel like we were doing something, and sometimes that is more important than anything else.

Now, the first thing most sailors would be wary of when riding into the clutches of a potential maelstrom is the risk of de-masting, i.e. the possibility that either wind or waves might rip the sail clean off the boat, taking the mast itself with it. Needless to say, this is bad news. As a sailor anywhere, let alone in the middle of the ocean, losing the mast is guaranteed to wreck your whole day. Not only does a de-masting rob you of your main source of propulsion, but it also makes you very unstable, and if the blob of aluminium, cables and canvas that used to be your rigging doesn't completely separate from the hull it can do a pretty good job of pulling the rest of the boat over and sinking you altogether. So, a pretty shitty scenario altogether. We were quite keen that this shouldn't happen to us. However, we were equally set on ensuring that if it *did* happen we were as prepared as we could be. So, we conceived, designed and manufactured a very specialist tool to help out in the event that the worst should come to pass. We christened it, and you must excuse my language: The Mast F**ker 2000!

The MF2000 consisted of a large set of bolt cutters, a hack saw and a sturdy hammer all tied very securely together, albeit only with string, forming one rigid tool. This piece of advanced hardware was to be stored next to the companionway AT ALL TIMES so that in the event of the mast breaking away we could grab it, jump out on deck and clip, saw and bash away whatever umbilical cord of cable or twisted metal remained, casting the wreckage off and freeing up Gambo's superstructure from the giant sea anchor of death that used to be our friend. Quite imaginative, don't you think?

We were all quite proud of the MF2000 and in my head at least it was quickly anthropomorphised and assumed similar crew status to that which Geoffrey already boasted. Perhaps this upset Geoffrey. Or, perhaps Sam is cursed, for as if in response to his precautionary 'what to do in the event of autopilot failure' guide penned in the log above, as Sam describes:

Shortly after preparing the para anchor system at around 15:00 the steering system failed.

Amazing.

I'm sure I've already described the continuous and pervasive sound of Geoffrey's hydraulics and the way that they both infuriate and calm you simultaneously. We hated them, especially when using the 'skipper's berth', but we loved them too, as they meant everything was working. Well, as Sam says, after we had undertaken all the ship-saving measures I've just described, Geoffrey went AWOL. I remember the whining of the hydraulics assuming a more rapid tempo, and the sound of the sails suddenly beginning to flap. Then there came the incessant beeping which meant that Geoffrey had really lost his s**t. I remember Matt's face: flat, confounded concern.

Panic stations. We hit 'manual' on the GPS display, silencing the alarm, and rushed up topside to assume steering stations in the building winds. Sam describes what happened next in a log entry:

Both the autopilot and the steering wheel were unresponsive. We were bobbing clumsily and out of control on an ocean full of wind and devoid of landmarks. Suddenly the sea had taken charge of us and was tossing us around as a dog might play with its favourite chew toy. Without steering we could neither turn nor hold a steady course to the wind.

Colin and Matt quickly installed the emergency tiller (a metal pole which inserted straight into the top of the rudder, bypassing the main steering wheel) and Matt started manoeuvring us. Soon however he realised that something was 'banging'. The tiller would only go so far in one direction.

To investigate we ragged out the contents of the port side locker and found that the universal joint holding the steering arm had come apart.

Colin grabbed the tools and I crawled upside-down into the locker clad in old clothes from Alun's locker. The offending arm was soon bolted back on and appeared to be working fine. Of concern is the tapping valve on the top of the hydraulic ram which may be bent. It does not seem to be leaking and the AP and wheel are working fine now. Could inspect it in a day or so, if the weather is good.

Read between the lines from that last statement…. "*Could inspect it in a day or so, if the weather is good.*" This, in Sam's usual understated fashion, is a coded way of saying that he was pretty ill after performing that repair. I would have been in absolute pieces! The man was upside down, in a diesel and oil soaked storage locker measuring approximately a foot-and-a-half in width and the same in depth, wearing a head torch, repairing a fairly technical piece of hardware on the now swelling waters of the North Atlantic. Ten points old chap! I'm actually feeling slightly queasy just writing about it now.

Even by Sam's definition I think this impromptu repair was 'a bit of fun', and he was justly proud of himself, even deeming to draw some diagrams of the repair job in the corresponding log entry. Sam has not yet produced offspring, but perhaps one day he will. I will therefore include one of these diagrams for posterity:

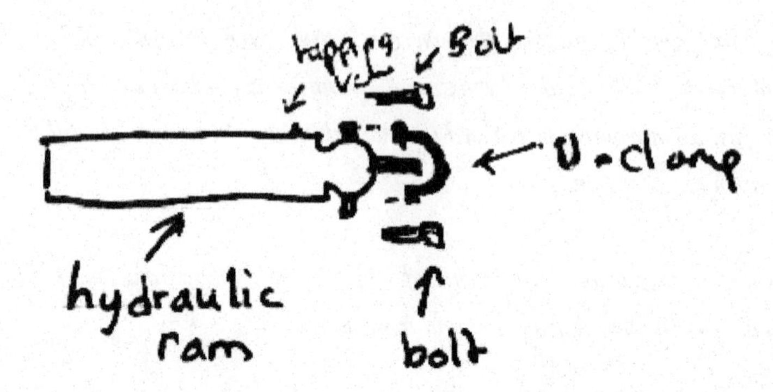

Thank you Sam! Our newly repaired steering mechanism meant that we were now able to steer a more direct course *into* the approaching storm.

You'll recall Sam's log from a little after midnight on the 22nd. If not, it read as follows. I've taken the liberty of underlining the pertinent section:

The hurricane is forecast to the south of Newfoundland, Canada today, and tracking up to the SW coast of Greenland tomorrow. Fortunately we are quite far east and may miss the 50kt winds.

Well, before the sun had completed another full revolution in the sky, Sam put pen to paper again, by which point things had changed somewhat. It seems that, unsurprisingly, our mates back in the real world had access to better weather information than we did, and the winds came in considerably stronger than the 35 knots which Sam's records show the GRIB file had suggested we might face:

22/09/10
Time not marked. Sam:

We've seen Force 10 now. It's been continuously Force 8 for some time. It's difficult to write. I'm constantly aware that this boat doesn't float. It's just temporarily displacing the water around it. I've watched the tin cans we throw overboard – they stay on the surface a while, before sinking. 36kt. 37Kt, 39kt, 41kt, 38kt.... Matts making pizza.

As it turned out, unbeknownst to us at the time, this storm even had a name. It was called Hurricane Igor – which I think sounds suitably nasty – and it actually proved to be the strongest hurricane ever to have hit Newfoundland since record keeping began. As this system completed its long journey from the warm waters off western Africa to the Arctic depths off Canada's east coast, wind speeds over parts of Newfoundland built to in excess of 100 miles per hour.

We were lucky to be a bit further east by this time. At its peak, the wind that we on Gambo experienced during this period breached only 55 knots, equivalent to a little over 63 miles per hour. This actually put our 'gale', which Sam notes as having tipped into the realms of a 'Force 10', more into the realms of a Force 11 at times, according to the Beaufort scale. But really, when you're in a small boat like we were, the distinction between a Force 10 'whole gale' and a Force 11 'violent storm' is somewhat academic. What I can tell you for sure is that it was unpleasant.

Everything becomes hard work, and its tribute to Sam's sea legs at this stage that he was able to write any log entries at all. (You might have noticed that neither Matt nor I have contributed any in a wee while now.)

Sam's comments about the boat merely *"temporarily displacing the water around it"* really strikes a chord in me, and his description of how tin cans behave when you watch them in the water is bang on the money. We were all constantly aware of how precarious our situation was. We were a steel boat occupying that transitional zone between air and water, our yacht being the only life-supporting island of habitability for hundreds of miles in every direction.

Steel does not float. And we were far from watertight. As we've already discussed, the ocean is perpetually finding a way into your living space. It comes in on your clothes. It comes in through cracks in the hatch seals. It finds a way into the bilge through the anchor fairlead on the bow, especially when the boat is forever digging into the type of water walls which a Force 10 will throw at you. Basically, you are perpetually sinking to a lesser or greater extent. If the volume of water inside reaches a certain threshold, you lose buoyancy and it's all over. Of course we had a bilge pump with which to empty the pooling ocean from the bowels of the boat, but this, like everything else, could break, and I can tell you that with the worrying regularity with which we were having to use said pump at this time I was, perhaps unwarrantedly, concerned for its wellbeing. Bailing with buckets would have done nothing at all for our morale. Fortunately though it never came to that.

Food on the other hand does EVERYTHING for morale, and somehow, as Sam recorded in his log, despite almost perpetual seasickness, Matt managed to muster the energy to make pizza. Pizza! From scratch! Phenomenal. Things like this are what get you through. It is escapism, pure and simple. If there's something complex like pizza on the menu then everything must be OK. Right? Instant noodles on the other hand, although welcome every time, signal that things truly are bad. When noodles come out it's because no-one aboard has the faculties remaining to perform any galley-related task more complicated than simply boiling water. Like the Ministry of Sound, Chillout album, pizza is a sanity saver and I will never take it for granted again.

So, we weathered the storm as best we could. In truth there are times when this is genuinely good fun. During the day you can see your enemy and look him right in the eye, and I've had some truly cathartic experiences spent alone on day watch during times of severe weather. When the wind whips every inch of your body and you are helming the boat manually you feel like you are truly riding the world. The sea is a serpent, the wind its wildness, and you are a warrior riding this treacherous beast into battle. (See, that's the Advanced Dungeons and Dragons player in me coming through).

Truly though, I have always revelled in wilderness, challenging environments and extreme conditions, and there is nowhere on Earth, save perhaps for the Polar Regions (which, remember, we had just come from), more wild, challenging and prone to extremity than the open ocean. It teaches you about yourself. Sometimes a little too much about yourself, truth be told; visceral winds can make you feel alive like nothing else can, and few things put life in perspective like staring up on a towering wave as it bears down on your little floating world. These are all experiences for which there is no substitute, and when you feel that inside your heart there is simply nowhere else you would rather be than right there on that boat in the clutches of a gale. Night time, on the other hand.... That's a different story.

Storms at night are the stuff which fear is made from. Wind shrieks out of the darkness like best forgotten memories and the waves crash against you unseen until the last moment, every one a shock to your system. There are endless places for nightmares to hide in the three-dimensional misery that is a night time storm. You feel like an island of light and warmth, surrounded by invisible forces which are trying to destroy you.

We were all tired again by now. Perpetually tired. Although it's amazing how much time you can spend sleeping during a storm the quality of that sleep is very low. Your body switches off, and you lose consciousness, but as the elements batter the boat and your body is tossed every which direction some part of you is kept awake and alert, probably through simple self-preservation. Sleep during a storm is a shallow, dreamless affair. So, when you wake you rarely feel truly rested. This has a pronounced effect on your brain, making trifling matters seem difficult or even insurmountable. This is one reason why long-distance sailing can, at least for me, be extraordinarily difficult if your mind, or heart, is elsewhere.

16 / Disaster

It isn't only psychological fatigue which can take its toll. The physical challenges of prolonged sailing, especially in severe weather, are far from insignificant. We were all being worn down by the constant battering we were taking, by the effects of being perpetually wet and from the physical toll of working on deck in adverse conditions. Storms are changeable beasts, requiring constant attention to the sails. So, we were regularly required to don foulies, grab a harness and venture out into no-man's land and either reef or un-reef a sail. I was scared every time I had to do this in rough weather, but partially because of this fact and partially because I genuinely did know Gambo's rigging better than either Matt or Sam, I made a point of spending more time up on deck than perhaps I needed to. I've never been the beefiest of blokes, and this was taking its toll. My chest muscles were becoming strained from the effort of winching Gambo's colossal, main sail up, time after time. As you can imagine this is particularly trying at night, when ropes fly unseen about your head and only the spot directly ahead of you is illuminated by the narrow beam of your head torch. I remember having one particularly bad time on a different voyage, several years previously. On this occasion, we were in Antarctic waters, heading north from the west coast of the Antarctic Peninsula to the Falkland Islands. It was pitch black and we were being battered by a gale and, it being the Antarctic, it was the kind of gale that throws not only air and water at you but ice also. The snow was driving hard and the darkness was full of frozen threats, both airborne and floating in the sea. Every gust of wind which barrelled out of the night served as a reminder that here was not a good place to mess up.

I had been 'deployed' on deck to reef the mainsail, alone. So, with foulies pulled tight, safety harness done up and head torch on I ventured out to tame the beast. I don't remember how strong the winds were, but I can tell you that the fiercest airs we sailed in that season reached 65 knots – a solid Force 12 'hurricane' on the Beaufort scale. The Southern Ocean doesn't pull its punches!

I un-cleated the main halyard and lowered the canvas to the level of the appropriate clew, at which point I hooked it over the goose neck on the mast and reached for the halyard once again in order to winch the sail up and tight again. As I did so, I somehow fumbled and released the rope into the wind, likely set off balance by the impact of a sudden, unseen wave. As soon as it was free the rope came alive. Such was the force of the wind that the errant line lashed around in the air like a bull whip, and in the darkness it was all but invisible. It was scary, and I would duck and dodge as the rope intermittently whipped through the beam of my torch and past my face. I grasped into the inky air, flailing to retrieve the all-important line as the mainsail flapped and bellowed boorishly next to me, it's folded, frustrated expanses thrashing in the gale with only my right arm to restrain it until I could tighten the halyard and trim the reefing lines.

Suddenly, the airborne rope cracked in the night and whipped me in the face. It hit me with such force that it not only knocked the torch off my forehead but it also opened the battery compartment, sending three triple A's flying into the night. I felt like I had been smashed with a nightstick! Suddenly I was plunged into darkness, leaving me clutching the flapping sail with one hand and my face with the other as the rope snake continued slashing the space over my head. There was nothing I could do without light, so I crawled back aft on my hands and knees, leaving the sail to its raucous protestations.

I managed to borrow another head torch and was able to get back out and complete the job, even finding my torch in the process. It was sitting in the gunnel being swept by the invading ocean. The batteries on the other hand were long gone. I hadn't been injured either, much to my pleasure, as fortunately the torch had taken the brunt of the rope's blow and I had only a small bruise in my eye to show for the episode. That said however, I didn't forget the experience in a hurry and certainly took that little bit more care with ropes at night from then on.

Fortunately we had no midnight rope regains to deal with during this second storm in the North Atlantic. Our rope-related calamity this time was of a different and far more serious nature. Sometime early in the afternoon of the 23rd of September the wind was abruptly knocked out of our sails, so to speak:

23/09/10
1700 UTC. Sam:

The blue + white furling line for the jib snapped and the jib got let out to its full extent. We couldn't save it.

'And that's all he wrote'. I think it's what Sam *doesn't* say which speaks loudest in this log entry.

By early afternoon of the 23rd we were beginning to think that we were through the worst of the storm, and thus far no damage or injury. Things were looking up, and although the wind was still blowing a stiff 35 knots and the seas were still kicking up a relatively severe 3 meters, we were optimistic. It was daytime after all, the ship hadn't turned upside-down, and we hadn't even needed to use the MF2000. I'll admit, I was perhaps a little disappointed that we hadn't needed to summon the MF2000 for *something*, but still, our persistent habit of staying both afloat and alive was a good result.

It was Matt's watch and we were running well with the weather, squeezing out a respectable 6.5 to 7 knots, if memory serves. Both Sam and I were getting our heads down. The mainsail was lashed on the third reef and the genoa was mostly furled so that perhaps only a metre-and-a-half of the sail's foot protruded. This wasn't much, but still it was sufficient to provide the all-important lift which carried our bow up and over the waves and permitted us to sail close-hauled into the easterly winds. Things could be worse I thought, as I lay curled up in my berth, which on that particular day I remember very clearly was on the starboard side of the main cabin, that also being the uphill side as we were on a starboard tack.

Yes, things could indeed have been worse.

And then, very suddenly, they were.

I wasn't exactly sleeping. More like dozing as we rode into and then out of each watery trough. I don't know if it was on an upward or a downward stroke, but all of a sudden I heard an abrupt BANG through the deck above my head and the boat heeled VERY uncomfortably to port. This tilt would likely have tipped me out of my bed despite the leeboard, that is, if I hadn't already been halfway out of my sleeping bag and on my way to the companionway. I stepped out of the berth and onto the table. Not onto the surface of the table mind you, but rather onto the edge of the table top, as the boat was now momentarily at something approaching a 45 or 50 degree angle. I had no idea what had happened, but it sounded bad.

Within moments Matt's face appeared at the companionway: "Guys!"

A simple but entirely adequate rallying call.

Sam was busy doing exactly what I was, only he was in the position of scrambling uphill rather than downhill from his berth on the port side, and so gravity wasn't on his side. To be fair though, at this precise moment gravity wasn't on ANY of our sides.

Sam and I raced up on deck immediately, ultimately foregoing the usual foulies. I had a feeling that whatever was going on, we were going to get irretrievably wet regardless of what we wore. I wasn't wrong.

We emerged from the hatch to see the genoa FULLY unfurled, its whole double-decker, bus-sized canvas was exposed to the gale. That's why we were suddenly leaning over so far!

As things happened, by the time we were up on deck Gambo's tilt had become more manageable as the genoa, what with not being sheeted in, was 'sailing' very inefficiently. Air was spilling from its slack face and it was luffing energetically. So, without really thinking about it in any great detail, I reached over to simply furl her back in again. To my initial confusion and subsequent dismay I found the furling line VERY easy to pull in. Too easy in fact, the reason being that it was no longer attached to anything. As I drew the rope in from hand to hand I came suddenly to a ragged end. The line had snapped.

We'll never know for sure whether the rope had chafed or whether it simply reached that time in its life and snapped on account of 'natural causes', but what we later decided was most likely is that what had initially been meant as just a bit of fun had ended up stinging us badly. Before leaving Nuuk, Tom Cowton had lashed a caribou skull to Gambo's bowsprit, intending it to be something of a gruesome figurehead. It's possible that at some stage – probably as we played with the sails during the night – the genoa's furling line had caught around one of our figurehead's antlers and, over the course of the last twelve hours or so, had chafed through. The bang I had heard from my berth was it finally giving up the ghost.

This wasn't good. All however was not lost! We decided very rapidly, and in retrospect perhaps rashly, that the thing to do was to drop the genoa altogether and stow it on deck. Then we could raise the staysail and deal with replacing the rope at a later time when the weather was better. We just needed to get that massive slap of canvas out of these winds, and quickly. A piece of fabric that size roaring loose in a 35 knot wind strikes fear into the heart and we wanted it gone. So, Matt quickly started the engine and relieved Geoffrey, taking manual control of Gambo in order to point us into the prevailing weather. I raced up to the bow and VERY carefully, with a diving dagger clenched firmly in my teeth (it makes sense to me now why pirates do this!) crawled under the railings and onto the bowsprit from where I could haul the genoa down and off of its runner. Sam moved quickly to the port side deck where the sail would be coming down in order that he could bring the canvas onto the deck as it fell, controlling it by hand and therefore preventing it from trailing in the water. (This all sounds great in theory doesn't it?!)

Despite the conditions, I soon had the small retaining shackle at the base of the forestay runner unscrewed and was ready to pull on the canvas. All the while Gambo was riding up and down on the waves which, as we had intended, were now taking us directly on the bow. The idea of this approach was to prevent the sails from catching any wind as we worked with them, and it was indeed having the desired effect. However, it was also causing me to be intermittently lifted high above the water – perhaps 3 or 4 meters clear – only to be plunged back down again and submerged, often up to a metre deep, in the belligerent sea as Gambo rode violently through the spume. I was clinging on for dear life, and, I think, stupidly doing it all without a safety harness. My memory is sketchy on that one. Sam reckons he made me put one on, which is reassuring. What I do remember exquisitely vividly though is that at one moment I would be underwater and holding my breath, while at the next instant I'd feel a lurch, opening my eyes only to see that the water was now frighteningly far below me, knowing full well that in another second it'd be rushing back up to meet me.

The strain was tremendous. I was being ripped alternatingly upwards and downwards by the force of the momentum. It was only during the moments above water that I could try and do anything useful, but after every sudden dunk in the ocean I'd be momentarily incapacitated by salt water. Every few seconds my eyes, ears and mouth would be filled with the sea. Every time Gambo's lurching bow hoisted me clear of the ocean I'd spit out the salt, wipe my eyes with one hand and tug on the canvas, straining to coax it down the forestay.

I remember Sam shouting: *"You're my hero Colin!"*

He was, of course, laughing, which is fair I'd say.

As I was repeatedly being dunked on the bow, Sam unfastened the genoa's halyard rope, allowing it to run freely, and I began pulling down and stowing the sail by my knees. This caused me some difficulty. The sail, unlike the ropes, was almost brand new, and the canvas was thick and stiff. It didn't much care for being dropped it seemed. But, I put my full weight into it and, timing bursts of aggression with the lowermost points of my reliably timed immersions, I managed to bring momentum and gravity onto my side and gradually coaxed the sail down and out of the wind. Meanwhile Sam bravely reached over Gambo's guard rails and brought handful after handful of the big white monster onto the deck and out of the wind's clutches. Suddenly however the sail jammed. It wouldn't move. I tugged and tugged, but it just wouldn't give. We had still only succeeded in dropping perhaps two thirds of the genoa's overall length and a sizeable portion was still flapping defiantly above us. I looked back deckwards through a wall of spray and saw where the problem was. The spinlock mechanism, a one-way, ratchet-type device designed to prevent ropes from slipping away from you under strain while you pull on them, and through which the halyard was threaded, had gradually engaged itself as the rope had been sliding through its teeth. Now it was seized. All that needed to be done was for someone to reach over and flick a lever and all would be well. But who was there to perform this simple task? There were only three of us! Matt was back in the cockpit steering and couldn't abandon his post. He was fixed to the spot holding Gambo in the storm, forced to watch us struggling up front which, I think, was probably more uncomfortable than actually being there. I was locked rigidly between the bowsprit and a railing, holding on for dear life as the sea battered me, while Sam was standing a mere 2 meters away from the mast. It seemed plain that Sam was the best placed to sort things out, and I shouted to him over the wind: *"Sam! Hit the spinlock! The rope is jammed!"*

This was a mistake. I should have gone. At that point, although none of us realised it yet, Sam's existing job was by far the most important. Sam stepped mastwards, stretching to reach as far as he could without abandoning the canvas. He necessarily let go of the sail with one hand in order to hit the appropriate lever, and at that second a gust of wind leapt up, spilled over the deck, seized the bulk of the sail from Sam's other hand and plunged it into the water. As Gambo was of course moving forwards under engine the genoa quickly trailed aft, guided by the tremendous forces of both water and wind. Within moments it had slipped under the hull, found the propeller and wrapped itself resolutely around it. We called to Matt – who couldn't really see what was happening – to put the gears into neutral, but we were literally a second too late. The boat's engine stopped with a thud.

And things had been going so well too!

For a long few moments we stood in our respective places, shocked. Had this really just happened? Had we really just wrapped the boat's largest sail around the propeller? Yes. Yes we had.

S**t.

As Matt's job was now officially over, he left the helm, first switched off the engine and then came up to where Sam was standing. I too crawled out from my watery perch and joined them. It was easier now, as the boat most certainly was not moving forwards into the waves.

Together we just stared at the vast tangle of canvas and rope which was now pulled tight over the guard rails and into the thrashing water. What now?

The drag caused by this colossal blob of sail had brought us to an abrupt halt. We were now quite literally adrift in the waves, bobbing about like a helpless cork. Honestly though, this was the least of our worries. We wanted both our sail and our engine back. Quite naturally, the first thing it seemed prudent to try was to pull on the confounded thing with all our might. So, that's what we did. Repeatedly.

This didn't work. The sail held fast to the propeller.

Next, although perhaps it was a stupid and forlorn hope in retrospect, we tried starting the engine in reverse whilst pulling simultaneously. A good plan in theory. But, the sounds which this move created in the engine compartment VERY quickly persuaded us that this was a BAD idea. In retrospect I shudder that we even attempted this, for if we had sheared the prop shaft the drag from the sail could well have sent the tail end of said shaft slipping happily out the stern of the boat, leaving a nice 3-centimetre diameter hole which would quite quickly have sunk us.

Next we tried turning the propeller by hand, placing the engine in neutral and manually rotating the transmission which was located under a floor panel just beneath the companionway. Again, no joy. The propeller must be VERY well clogged as it wouldn't budge. We really were in a bit of a fix it seemed.

With mounting frustration, and perhaps a reasonable amount of panic beginning to set in, we decided to clear as much sail as possible from the propeller in the hope that perhaps we would be able to make reasonable way without the genoa. After all, we really didn't need the engine right now. Not in the open ocean. So, it came down to simple hacking and slashing. In something of a proactive rage we all grabbed knives, hack saws and the boat hook and attacked whatever we could snag and haul up within arm's reach in an effort to loose the massive sea anchor which was wrapping Gambo's port side. I quickly cut the string which was binding the port guard rails, loosening the whole cat's cradle of cord and canvas and allowing us to reach that little bit lower.

Together we eventually managed to hack our way through as much of the sail as we could reach, freeing the lion's share which was still attached to the forestay. The remainder slipped slowly from sight into the water, trailing aftwards. We moved to the stern and, to our dismay, saw that a rather huge amount was still in the water. Gambo's propeller was situated a good 2 meters forward of the transom, and there was at least a meter-and-a-half of trailing canvas visible from where we stood at the stern railings. That amounted to a hearty 3.5 meters of white, wobbly drag.

We were a sorry sight: no engine; a large octopus of debris trailing from our prop; the guard lines on the port side lying severed on the deck; the carcass of the genoa now flapping limply from the forestay.... We were ugly, and very unhappy.

In an effort to tidy up a little I went to clear away the remains of the genoa, still attached to the forestay while Sam and Matt began fishing off the stern; one with the boat hook and the other with a knife tied firmly onto a length of pole. It was a fairly desperate effort, but anything we could do to reduce drag would be an investment of time. I went to bundle up the dead sail on the bow, but after undoing the shackle which attached the genoa to the halyard with some difficulty, I found the cord which had been used to tie the lower end to the base of the forestay impossibly tightly tied. I just couldn't undo it, so in a frustrated rage I grabbed a knife and simply cut through it. Seriously, in the bigger picture at this stage, one more cut line wouldn't make anyone cry.

I also cut away the strings holding the caribou skull in place on the bow and, in a belligerent fury, hurled it full-pelt into the sea. Needless to say, no-one complained when I later announced that I had done this.

We all worked as best we could to clear canvas from the water behind us. For several hours we kept at it, maybe just unwilling to accept that we were fast running out of options. It wasn't until it started getting dark that we gave up:

23/09/10
Time illegible. Sam:

@ 2008Z we finished trying to free it (the sail) *from the prop. N59° 28.6057' W28° 31.8867'.*

Disgusted and depressed we all went below for a while. I think we maybe had a cup of tea, or several. I can't remember. It seemed virtually pointless keeping a watch as we were essentially hove-to, but we had recently read a line in one of the numerous sailing books aboard that it can take only twenty minutes to enter a mid-ocean collision situation. This, given the average cruising speed of large ocean-going vessels, is the time needed to go from having a clear horizon to being run over by a cargo ship or similar. Once you know this, you never treat the horizon the same way again. So, even though we ourselves were going virtually nowhere, we kept up with the watch rota.

17 / 'The wettest man; let him go to the well'

With the remains of the sail stashed below deck, and the guard lines tied away onto the stern railings, we now at least *looked* a little better. Gambo still however wasn't going anywhere quickly. We set and sheeted the mainsail again in an attempt to make some headway but, sailing into an easterly wind without a headsail of some kind, we were unable to make anything even resembling a useful course home; and even on a useless bearing of almost directly north or directly south, Gambo could only achieve about 1.5 knots. This wasn't too inspiring.

We deployed the staysail, but still couldn't make headway into the weather, and still our maximum speed seemed restricted to about 2 knots thanks to the very unwelcome passenger trailing in the water behind us.

We decided, given our inability to sail even an unflattering caricature of an easterly course, that we'd best drift north. We were in fact only 300 nautical miles from Iceland. So, perhaps salvation could be found in Reykjavik?

Around this time Sam recorded a short video on his camera. It gives a little insight into how we felt at this point in the proceedings.

Sam (sounding tired and miserable): *"Where are we Matt?"*

Matt (smiling): *"Erm, about 250 miles south of Iceland I think… It's amazing how fast it's gone from…like…we might get home pretty quickly to lets, err…let's try and get home, eventually, at some point [nervous laughter]."*

Sam: *"Do you want to say anything to your lecturers in Bangor?"*

Matt: *"Oh, I might have to start next year at this rate [laughs]. I'll have to defer my course."*

Sam: *"We're not really moving anywhere…"*

Matt: *"It feels colder around here…. I want to go south. Shall we go south? We could just go south for a long way, like France."*

Sam: *"Spain would be nice at this time of year. I've heard deep-water soloing in Majorca is quite good."*

Alun had in fact told us, quite explicitly, that we were to avoid all ports outwith the UK, i.e. that we weren't to stop anywhere en route to Oban. I had no inkling as to why this was so important, but then leaders/managers/bosses/boat owners don't always feel the need to explain themselves. These were his rules. But then, he wasn't the one stranded in the mid-ocean, at least not on *this* occasion! So, I'll admit I spent quite a bit of time reading the sections of the pilot book which described Icelandic ports and harbours over the next twenty-four hours, especially the pages on Reykjavik. I had no desire to spend the next two months at sea! We worked out that, given our current position and distance from home, even assuming a solid westerly wind for the whole run (spectacularly unlikely) at our current speed it would take us something like another seventeen 'good sailing' days to cover the remaining 800 or so nautical miles to the UK. So, in reality, given that in our current state we had very little control over our course, we were probably looking at something more in the region of one or two months. To say that this prospect caused some amount of negativity aboard would be an understatement. So we resolutely, albeit privately, maintained a course for Iceland. This might yet well have proven to be our only option.

However, as luck would have it, sometime during the morning of the next day and after what now amounted to over twelve hours of dejectedly drifting, an alternative solution presented itself. Deep in the kit pile which was our on-board lavatory we had located what could potentially be some useful pieces of equipment. These were:

1: An air compressor

2: A scuba diving cylinder

3: A scuba jacket for regulating buoyancy

4: A breathing regulator and diving mask

We (or at least I) had previously not know that these bits of kit were even there! I suppose if we had been stopped at 'customs' prior to sailing from Greenland, and anyone at 'baggage check-in' had asked: *"Did sirs pack their own boat?"* we would've had to have lied.

So, it appeared that there might potentially be a way of getting under the boat and sacking off our big white, canvassy show stopper after all! The only problem: none of us were scuba divers.

Our collective scuba diving curriculum vitae read something like this:

Colin Souness:

Aged fifteen-and-a-half years, Colin once undertook an introductory dive session in Orkney which is on top of Scotland. It is very cold there but his mummy thinks it is very pretty. He had a jolly good time and, despite showing complete ineptitude at controlling his buoyancy, thoroughly enjoyed the novelty of breathing underwater and seeing some very pretty shipwrecks in calm, shallow water.

Colin likes fish because they are pretty and because they taste good.

Time elapsed since last dive: eighteen years.

Samuel Doyle:

Sam has never scuba dived before but he really likes a nice swim and thinks the sea is really fun. He doesn't mind getting wet. Sometimes Sam gets wet on mountains. Sometimes Sam gets wet in caves. Sam is friends with Colin and Matt because they don't mind getting wet in silly places either. Recently Sam has mostly been getting wet at sea. Getting wet is fun.

Time elapsed since last dive: ∞ *(infinity)*

Matthew Burdekin:

Matt likes sailing. Sailing is fun. He likes it even though it makes him feel spewey. Sailing makes him happy, but then Matt sometimes likes having a bad time. Matt is friends with Sam and Colin because neither of them mind getting wet on mountains and things. They like having bad times together.

Matt once had fun doing an introductory PADI dive too. He liked it. It was nice.

Time elapsed since last dive: about two years.

So, although there was now an option, it was not one we were going to explore lightly. A crippled boat, 300 miles from land in the middle of the North Atlantic, wasn't exactly the safest place to learn how to scuba dive. And, as far as tasks to complete prior to mastering even the most basic diving skills, climbing under a perpetually moving vessel with sharp implements in hand and a spider's web of rope and canvas to kill has got to be one of the less appealing. But, such was our situation. I have a very clear memory indeed of all three of us sitting in the cockpit at the aft of the boat, staring at the amassed diving gear which was sitting in a pile on the deck between us. We were very quiet.

The compressor looked rusty. The fittings on the cylinder were slightly corroded and their plastic valves had been broken off, requiring that a pair of pliers be used to open and close them. But, at least the regulator and the mask looked in good condition!

We sat around the gear, staring at it in silence.

Eventually, Sam ventured to speak: *"Are we really going to do this?"*

I spoke next, perhaps only a few seconds later but after what felt like a long pause indeed: *"I don't think we really have much of a choice."*

There followed another equally uncomfortable, but probably surprisingly brief, gap in conversation. Then Sam spoke again, and said what I think will always be, for me at least, the most memorable quote of the entire voyage.

"That's committed boys."

He was absolutely right. It *was* committed. But who would we commit? After a surprisingly brief review of our respective scuba résumés there seemed to be only one 'sensible' candidate: Matt.

As our resident scuba 'expert', Matt, never one to choke at the prospect of a good challenge, stepped up, just as he had done on that fateful helipad back in Uummannaq almost a month previously. The game was on.

So, firstly the tank needed filling with air. We scrubbed it down and checked that the regulator at least fitted onto it. It did! This was excellent news. Next, we needed to get the compressor running. Now, the compressor was quite a characterful beast. Hailing from somewhere in South America (exact nationality unknown) it had seen many a season at sea – including several expeditions to the Antarctic – and over the course of its adventures had likely spent more time soaked in salt water than most of us do in a lifetime. How long it had sat idle in the bowels of Gambo's 'shower cubicle' was anyone's guess. Suffice to say that I think it had been a very long time since anyone had attempted to start it. Perhaps I'm wrong, but it certainly took a good while to shake the beast into life. This onerous (but potentially very satisfying) job fell to Sam who was by far the most capable in these matters. After what I remember being some time, perhaps forty-five minutes (Sam remembers it being hours…), we all heard the bronchial spluttering which asserted that he had succeeded. The compressor was alive! It perhaps wasn't happy, but then I suspect I wouldn't be either if I had just been asleep for several years. Especially not if I were to wake up only to find myself stuck on a handicapped boat in the middle of the North Atlantic.

We duly hooked the compressor up to the dive cylinder and waited as the somewhat weary petrol-powered air squasher pumped breathing juice into the battered tank.

After what may have been about twenty minutes the pressure gauge on the tank seemed to stabilise at about one third of full capacity. We waited for another ten minutes or so, but it wouldn't budge. Of course we had no way of confirming whether or not this gauge even worked properly. But, we didn't want to waste petrol spilling air everywhere. So, we cut the motor and disconnected the tank. It seemed that we were at the stage of conducting a test dive!

While Matt donned a poorly-fitting dry suit (which actually contained a fair amount of water already) and put his kit together, Sam and I dropped the mainsail. Gambo was now dead in the water, or so we thought. Next, we had a think about how we could conduct 'diving operations' as safely as possible. We were already finding this crossing to be quite tricky with three people, so neither Sam nor I really fancied doing is as a duo. Besides, we both quite enjoyed Matt's company, despite the Johnny Cash, so it would be nice if we could avoid losing him. So, we settled on a reasonably straightforward safety system involving me going into the water as well, equipped with a dry suit, snorkel and mask, to keep visual contact with Matt (who would of course be a couple of meters forward of the stern and quite out of sight to anyone on deck). Sam would stay aboard where he would be in charge of two ropes, one of which would lead to Matt and the other of which would be tied around my waist. Not too shabby! We kitted and rigged ourselves up and took the plunge.

This first dive was a real learning experience on a number of counts. It was still relatively choppy and a stiff breeze was blowing. For starters this meant that the boat was actually still sailing, despite having no canvas up. The mast, rigging and hull still caught the wind, and as a result I suspect we were still making leeway at about 1 or 2 knots. Therefore, Matt couldn't simply swim over to the propeller after slipping into the water; he had to literally climb past the rudder and onto the screw against a slight current, and although 1 knot may not sound like much, I can assure you, after trying it myself a couple of times WITHOUT the scuba kit on my back, it is pretty tough going. It is, after all, the whole Atlantic Ocean which is working against you. This effort meant that Matt used air very quickly, and I think that over the course of the whole sub-aqua saga, he was never down for more than five or ten minutes on one tank charge.

The second factor which we hadn't thought of prior to getting stuck-in was that the boat was still rolling on the waves, meaning that as we clambered around beneath her the rather heavy and solid mass of her red anti-fouled belly was perpetually knocking us on the head. So, after our first exploratory submersion both Matt and I added climbing helmets to our rig. We looked pretty stupid, but the helmets were essential, and after all was said and done they were covered in streaks of red paint; testimony to the fact that they hadn't been for show alone.

I don't remember exactly how many separate dives Matt made before we were finished, but it was a lot. It was very tough going, especially as he had to do it all armed with a cutting tool of one kind or another, and he experimented with quite a few, including the legendary MF2000. This, of course, meant that he only had one free hand to climb or swim with against that current.

It was tough.

24/09/10
1100 UTC. Matt:

Waves were difficult and tiring. Most of sail removed, but still some of rope-threaded edge and small section of the sheet line left. Had to stop as knackered and was dangerous. Used a lot of air.

Current was strong enough to make regular free flow impossible.

Tried to see if engine would start in neutral. Did, but noise of banging so turned off quickly. If calm soon will (go down again and) *free the rest, but don't see it happening. Don't fancy going down again in similar conditions. We are sailing on now though, so will just have to deal with issue of landing* (referring to needing the engine to manoeuvre into a port) *later.*

Matt had done amazingly well, given the circumstances, but after several dives we were still clogged up. The drag had been reduced tremendously however, and we were able to make reasonable progress at a healthy 5 knots. But still, as Matt says above, without a genoa we couldn't sail effectively into the wind which, being an annoyingly persistent easterly, restricted us to going either NNE or directly south. We would NEVER get home on either of these courses! Fortunately however, despite Matt's understandable reluctance to go back under the boat given the conditions he had experienced on his previous attempts, this is exactly what he did.

The next day the wind actually let off, leaving us almost becalmed. Now, without any drive in our sails, we were left bobbing about without any options. We needed that engine!

With calmer waters and less of a leeway effect, Matt decided to give the prop another go. I remember his face, and it wasn't a happy one. Matt is one of the most positive people I know, and also one of the best and bravest rock climbers I've ever met. I have sat, perched on a relatively novice-grade rock climb, scared out of my wits, while only 10 meters or so to my side Matt was hovering about 150 feet off the ground, casually contemplating his next move on an 'Extreme' graded route. He doesn't get scared easily, but I know he was scared to get back in that water. Still, he did it anyway.

Fully re-equipped with dry suit, charged cylinder, mask, regulator and helmet, Matt slipped into the sea again, and I followed, once more taking up the comparatively comfortable position of spotter. This time he stayed down for much longer than before, likely because he was using less air in the improved conditions, but also because he did NOT want to have to go down again. Armed with a hack saw he attacked the fouled prop with gusto, determined that it wasn't going to beat him this time.

After what seemed like a long time I saw him slow his movements and stop. The blob of gangly sail and rope which had been so entrenched on the boats screw slowly started to move and, as I watched, it sank gradually downwards.

All of a sudden bubbles exploded from Matt's regulator. He wasn't in trouble though, he was cheering. He waved the hack saw above his head and gurgled with joy. It was a pretty memorable sight, and a memorable sound for that matter.

We both hung in the water for a few moments, watching the blob of sail sink gradually down into the ocean. It was genuinely one of the most beautiful things I have ever seen. The water was clear, and we could see to a depth of perhaps 10 meters. Faintly greenish rays of light probed down into the murky depths from above, cascading around each other as wavelets danced on the surface over our heads, scattering the sun's rays into the deep ocean. And into their midst the tattered wreck of what used to be our sail drifted: initially our means of propulsion but more recently a ball and chain around our feet. Now it vanished forever like a white ghost finally escaping the corporeal realm. It was spooky, and I will never forget that sight so long as I live.

We were free!

18 / Avante!

I cannot over-express how happy we were to be shot of our unwanted passenger. We felt like a weight had been lifted from us. And, in a sense, it had!

Nervously we half-turned the key in the engine's ignition. For six long seconds we waited with baited breath as the glow plugs heated, the circuitry beeping loudly to signal that this was indeed happening. Finally, after what seemed like an age, we turned the key the rest of its full revolution and.... Life! The cylinders fired, the boat shook slightly and the engine breathed with renewed vigour. It was working! We listened very closely for several minutes, none of us wanting to talk, checking for strange bangs, clanks, vibrations...anything which might betray damage sustained either from the initial entanglement or our subsequent attempts to start the motor whilst ensnared. But there was nothing. The old bastard seemed to be on good form.

Our relief was palpable. Suddenly, everything seemed achievable once more. The next check was to put the motor in gear. Had the prop shaft been bent? Had there been any damage to the carbon seal which prevented water entering Gambo along the shaft? We carefully engaged the forward drive and were happy to see that everything still appeared to be alright. We repeated the same check for reverse gear and still there seemed to be no issues. Water wasn't pouring into the boat through the stern; the prop shaft hadn't shot out the back of the boat in disgust.... Happy days!

This called for a celebration. As we had neither the urge nor the stomach for cigars or alcohol at any point during the voyage, we found other ways to get our kicks and decided that a swim was called for. Funny really, given how reluctant we had all been for anyone to get in the water only perhaps an hour before. Sam still has somewhat questionable video evidence, which he has shown in public on at least one occasion, of me jumping stark naked from the roof of the dodger and into the sea. We all did this a few times (not all at once, needless to say, so that at least one person would be left aboard at any given time), although I think I was the only one to go in the buff, and then I only did that once before realising that a camera had appeared.

Sam also had his first scuba diving lesson. It now seemed somehow unfair that while all of this had passed Sam had not yet experienced the singular joy of gazing upon Gambo's submerged belly. So Matt kitted him up and chased him overboard where he happily lolled for a good few minutes at least before the novelty appeared to wear off and he clambered back aboard.

As it turned out, Sam apparently wasn't the only one who was keen to check Gambo out at this time. No sooner had we all climbed back onto the boat and were getting ready to make way again than both a pod of pilot whales and a large fin whale passed right by us. To this day I regret not grabbing my diving mask and jumping straight back into the water. The view of that fin whale passing underneath our boat would have been unforgettable.

Anyway, in fairly short order we looked at our position, plotted a new course and got under-way (now once again under engine propulsion) in what we hoped would be the most tactically beneficial direction in light of the wind forecasts we'd seen. Looking at the predicted weather it seemed that there were more stiff blows in store, and that these would be another set of easterlies. What with the distance we had been pushed northwards off our intended track over the course of the previous few days of calamity we decided that Gambo needed some south 'in the bag', just in case we later found ourselves in the position of not being able to gain any ground in that direction. So, we re-started the engine in a flat calm sea and pointed Geoffrey toward 180 degrees on the compass:

25/09/10
1800 UTC. Colin:

A combination of bad wind, engine problems and sail problems has landed us 130 nm north of our course [interesting bias toward understatement in this entry!]. *Now we have engine we are heading south for ~12 hours, then SE for ~24 hours. This should help us to avoid a low pressure system and land us in the W/S winds and eventually make progress E/S for Rockall.*

Weather = 0 wind. Waves ~ 3 m.

And so we cruised on for almost twenty-four hours in virtually breathless conditions. After the tension of the previous few days this somewhat unexpected change in our state of affairs was extremely welcome, but also somewhat surreal. The reappearance in our lives of the engine's familiar hum had a very calming, almost soporific effect on us all. The wind, for the time being anyway, had nothing to say and we could go about our business without worrying too much about what was going on up on deck. There really wasn't anything going on up on deck at all! Those of us who were not on watch read, slept, cleaned a little and cooked again. Yes, instant noodles had made an appearance in our lives over the course of the recent storm, thus making it official that things had indeed been bad. But now those horrid harbingers of ill feeling were, along with the storm itself, just a memory. So, in the calm, flat-floored world of diesel propulsion, we baked bread and cooked up a curry. It was amazing.

Unfortunately however, things are rarely as straight-forward as they seem. I mentioned that the sound of the engine had a soporific effect on us, causing some amount of drowsiness (understandable after the battering both our bodies and minds had taken in recent days). Well, we soon became aware that perhaps it wasn't only the sound which was making us sleepy. A strange smell soon began to build in the cabin, and the air assumed an odd, slightly plastic taste. We didn't like this development one little bit and quickly stopped the engine so as to locate the source of the stink. Once again we dismantled the engine compartment which, although a fairly straight-forward procedure, quickly places the cabin into a state of complete disarray. The food lockers on top of the compartment have to be emptied, the dry foods and vegetable racks which usually sit mounted on one side of the compartment have to be relocated and the panels of the engine compartment itself all need to be temporarily stashed around the berthing area. It's a pain, I'll be honest, but the benefit of this engine arrangement is that once the compartment has been taken apart you have almost unimpeded access to the mechanism from almost every angle. You may not have much floor space left anywhere else, but still.... It's great for carrying out tweaks, services or minor repairs.

So we checked everywhere, both along the gas lines for the cooker and around the engine block, sniffing for any residual whiff and visually inspecting every valve, link, gasket and screw-fitting for any signs of a leak. We could see nothing. Everything appeared to be solid. There were no cheeky black streaks from the head gasket or the exhaust manifold. No visible holes and no pools of suspicious liquid. So, wondering if perhaps something had been spilled onto the engine from the food lockers overhead, something which had subsequently been boiled off or vaporised, we cleaned the engine off by hand using detergent and warm water, removing patches of dirt and grease the likes of which always somehow accumulate on even the most well-maintained go-box. We even replaced the alternator drive belt in case perhaps this fatigued looking rubber band had started to slip and burn. Happy that there now seemed to be no obvious issues we re-assembled the engine's housing box and started the monster up again. All seemed well for some time, and as it became dark we even sat down to watch a film on Sam's laptop. All we had however was Terry Gilliam's *The*

Imaginarium of Dr Parnassus – an odd film by anyone's definition and in any circumstances.

Somewhat worryingly, maybe about one third of the way into our little cinematic experience, the strange fumes began to build in the cabin once again. Unfortunately none of us were in the mood to do anything about it, and seeing as we were content that there were no structural issues or leaks which we should be worrying about we decided just to open the hatches, let the air flow through the boat, and ignore the issue for now. I've always had this somewhat silly notion that sometimes mechanical issues just fix themselves. I know its rubbish, but for some reason I still like to give things a chance to…well, heal I suppose. So, we decided to sit it out and see what happened. We were going to do our absolute best to enjoy this film and have a relatively normal evening goddammit!

Soon however, we all started to feel a little light-headed and, likely as a combined result of both its overwhelming oddness *and* the effects of the fumes, the film became almost unbearably strange. I in particular was having an especially unsettling time as I was on watch, so every fifteen minutes I would get up, leaving the movie for a second, and pop my head out of the companionway to check for other shipping. (Remember that twenty minute rule!)

Honestly, the combination of feeling woozy from the fumes, confused and disturbed by a weird movie (which we were watching in the dark and cosy berthing area) and then being reminded every fifteen minutes that I wasn't at home, nor in a cinema, but rather was watching this terrible film on a boat whilst chugging through the moonlit watery desert of the North Atlantic…. It was all too weird. None of us knew what to make of our situation right about then, and when the film finally ended we sat quietly for a few minutes feeling somehow bereft. The whole scenario was just too bizarre, and I have no desire to ever watch *The Imaginarium of Doctor Parnassus* ever again.

All of this being the case, the almost tangible gloom which that film brought down upon us did, somewhat unexpectedly, instil in us a very real need to improve our situation, and so as the feeling of bereavement slowly morphed into a proactive compulsion to somehow make things better, we resolved to fix the fumes issue.

Once again the engine compartment came apart and once again we got stuck in, this time with the help of the cabin lights and a couple head torches.

Eventually we decided that the problem most likely lay with a reinforced rubber water pipe which was touching the heat exchanger – part of the boat's coolant system. So, we attached it to another nearby pipe using a jubilee clip, lifting it clear of the scalding hot, heat exchanger. Once again the engine compartment was re-assembled and once again the cylinders were fired up, and this time…. No smell! Hurrah! This was a bit of a relief as now we could sleep soundly without fear of being poisoned.

The next day the seas were still calm and the winds still below useful levels so the engine stayed on. There was one very positive development that day however, and that was the arrival on-board of a replacement genoa.

"What's this?!" I hear you say! "How on Earth did you manage to arrange the delivery of a new foresail hundreds of miles out to sea?" Well, this is indeed a very good question. And the truthful answer is, I'm afraid, that we didn't. In fact, we'd had one aboard the whole time!

(– audible sigh –)

At Alun's suggestion, via text message, and in a fit of energy and pro-activity, Matt had decided to excavate the bow lockers to see what mysteries were lurking down there. To do this he had to shift a deflated zodiac (rubber power boat), a large, heavy and VERY expensive piece of bathymetric scanning equipment, various bags of field kit, skis…loads of stuff really. But, together we got into those lockers. And lo, a pile of white canvas lay folded in the bottom of the port-side storage bin. A sail! The moral of this whole story? If I had to settle on one it'd be that you should always, ALWAYS be fully in the know about what you have on board your boat before you sail anywhere. Knowing we had scuba gear and knowing that we had another genoa could have saved us at least two days at sea. But, then perhaps it wouldn't have been as much of an adventure?

Anyway, this canvassy gift from the gods wasn't a new sail really. Quite the opposite in fact. It was very old indeed, and had seen a lot of action already. I can testify to this, for I'd seen this sail six years earlier. We'd had some pretty big adventures together! It *was* a genoa, and I knew this from the arrangement of patches and repairs which it wore like marks of honour. The main clew – the metal ring which is stitched into the trailing end of any sail – had been ripped out and then re-attached using several thick lengths of strapping which had all been sewn on through very thick canvas. I know exactly how it was done because I was the one who had done it, sitting on the quayside in Ushuaia, Argentina, six years earlier in 2004.

So, while the relatively calm conditions persisted we ventured 'upstairs' to run the 'new' piece of canvas up the forestay. This actually proved to be something of an adventure in itself, at least for me.

As I undid the shackle which we had used to secure the hitherto useless genoa halyard onto the mast I clumsily let go of the rope. As the boat was still swaying around a little, and there was a bit of a breeze, the rope flew absent-mindedly away and out of my reach. This seemed to be fast becoming a habit of mine!

The halyard of course runs up to the top of the mast, into a small hole, down the inside of the mast and then out another small hole close to the starboard side winch. As the now liberated end of the halyard was swinging free off to port, the twin factors of relative lengths and the effects of the wind conspired to cause the 'business end' to shoot up the mast while the 'hands on' end came pouring out of the hole next to the winch, piling up on the deck near where I stood. This wasn't a useful turn of events, and I was left standing for a moment wondering why, oh why, I hadn't learnt over all these many weeks and months spent at sea, that you should never, NEVER let go of ropes! There was only one thing to be done: I was going to have to climb up the mast and retrieve the rope. Great.

Now, everyone has seen a yacht at some point. Most have probably been to a marina or a quayside where many yachts have sat berthed together. It's likely that most people who have, have also spent some amount of time gazing upwards at the collective mastheads, listening to the wind whistling amongst the rigging and occasionally catching the lines at the right angle and at the right speed to cause the halyards to clank and clatter against the masts. It's a unique sound, and in fact I've always wanted to live in a house from which I could hear this noise on a breezy evening. For me it's the sound of freedom. But anyway, if you have looked up at these masts, anywhere, you've likely thought to yourself: *"By jove, that's a trifle high isn't it!"* Or words to that effect anyway.

Gambo's mast was particularly high. She was a fifteen metre boat after all, requiring a solid fifteen metres of mast to accommodate a mainsail large enough to effectively drive her forward. Now imagine this fifteen metre mast being anchored at its base to a boat which was rolling around in the middle of the ocean, this roll pivoting around a point in space located in the centre of the hull. This means that as you travel up the mast, the amount of lateral movement caused by that natural ocean roll is exaggerated, so that in only a moderate swell the tip of Gambo's mast would be swinging back and forward to the tune of between 4 and 5 metres. That's quite a lot if you need to climb said mast!

As I'd watched the halyard shoot skywards I'd known what was coming. I knew I was going to have to get up there, and I knew I was going to have to do it now. This wasn't ideal. I'd climbed Gambo's mast quite a few times in the past, but only ever when the boat was safely out of the water and stood on a tarmac hard standing somewhere. Even then I had never climbed beyond the second set of spreaders. Thinking back, I'd been terrified even on those occasions. And now, here I was contemplating climbing to the very top, at sea, and in the midst of a not inconsiderable swell. I couldn't even face thinking about it. I was just going to have to do it. So, without saying a great deal either to myself or to the other two guys I climbed downstairs and grabbed myself a helmet, a climbing harness, a single sling and two carabiners. I told the lads where I was going and with Matt in tow behind me (just so at least someone would see if I were to fall) I headed for the base of the mast.

Now, generally you might want to be tied in, or for someone to belay you up to the top of the mast as you would on a top-roped climbing route. This would be sensible. Of course the best rope to use as a belay to the masthead would be...yes, that's right: the genoa halyard. Great. So, I decided, perhaps somewhat rashly, to do this thing without any belay, at least on the way up. I had a single sling which I attached to my harness at one end and, after tying a loop in the other end, wrapped it around my shoulders and clipped it back onto the waist strap of my harness. The sling would let me secure myself to the mast if I needed to stop at any point, either for a rest, to let a particularly rough set of waves pass, or just to man up and get a handle on my now growing adrenaline levels.

The good news, and there was some good news, was that Gambo's mast features rungs all the way up to the top, rather like some telegraph poles do. So, I really could just climb it like a ladder; a particularly high, exposed bouncy and swingy ladder perhaps, but a ladder nonetheless. But enough talking! I needed to climb!

I set off un-roped and shaking. As I climbed Matt stood below in the cockpit watching me and shouting. I think he was still suggesting methods whereby I could arrange some kind of belay, or perhaps he was already thinking ahead about something else, like how to bring the rope down. It didn't matter, because at this precise moment in time I just didn't want to hear him. After straining over the whistling of the wind in my ears, and stopping twice, with sweaty hands, to catch his words, I finally barked out that whatever he wanted to say, could he please just wait until I was down again. Maybe Matt didn't expect me to be so scared and uncomfortable – we were all climbers after all – or maybe he just had an absent-minded moment and didn't realise how distracting trying to understand him was for me. I don't know, and I still feel a bit bad for snapping at him from halfway up the mast, but I'm glad he stopped talking. I was trembling with fear, and needed to focus. Incidentally, after writing this book I sent it off to both Matt and Sam for checking. Sam sent back a very thoroughly annotated hard copy and by this paragraph he had reassuringly scribbled the words: *"He always does this."*

Onwards and upwards I climbed, stopping briefly on almost every second rung to regain my composure. I would thread one arm through the step on one side, around the mast and grasp the step on the other side, thus anchoring myself for a moment and allowing my heart a chance to slow down. I don't think I lost my composure at any point, and my face was probably just locked in an unhappy mask of general severity, but I was scared. And, the higher I got the harder things became. The sway of the mast was formidable as I neared the 'summit'. At the extent of each lateral swing the force on my body was terrific. It was like sitting on a trebuchet, or on some similarly massive catapult, gripping on with white knuckles so as not to be slung sideways and into the ocean. I daren't look down. When I did, and it only happened a couple of times, I not only felt the forces on my body but could see them too, which just made things worse. Looking down I would be staring into the water off to port of the boat one second, and then moments later would be peering off to starboard, Gambo's deck flitting from one side of my field of view to the other like a hypnotist's watch. It was horrible.

As I finally reached the uppermost rung of the ladder I locked my left arm around the mast again and with my right hand anchored myself temporarily using the sling which I had brought along. Now, with a feeling of moderate security on my side, I grabbed the end of the genoa halyard which, thanks to a nice, fat, spliced eyelet in its tip, hadn't shot through the hole in the mast and down into its inaccessible guts but had jammed on the outside. Using the second carabiner which I had brought along, I clipped the halyard onto my harness and closed the carabiner's locking screw-gate. Now I shouted to Matt – a little unruly-haired speck far below – that he could belay me as I climbed down using the other end of the halyard. In fairness, I really should have planned all this out and asked Matt to do this before I'd started to climb, but it was just one of those moments when I needed inertia on my side. Now, the job was almost done, and Matt understood me fine. He moved forwards, grabbed the bottom of the halyard, pulled the slack out of it and wound it halfway around a cleat on the foot of the mast for use as a makeshift belay plate. Now I detached myself from the masthead, chucked the sling over my shoulder, and began downwards as Matt gradually controlled the rope over and through the cleat as my weight pulled it upwards and through.

Perhaps only a minute or two later I was safely back on the deck, halyard in hand and my heart still racing.

Directly, we re-shackled the halyard line and attached it to the top of the replacement genoa. Still immersed in the inertia of 'doing stuff' I then ran her up the forestay. After only a short bout of activity the job was done and we stood back to admire her in all her fully-deployed glory. And quite the beauty she was too, sporting such a complex and, if I do say so myself, somewhat masterfully stitched array of patches. She was a great sight. In several places holes which ran right through the main sheet had been stitched over with small squares of fabric. They were a little scruffy (We did a few of them at sea!), but they would hold. Also, the sails leech line – which as we know is the thin cord which runs through the canvas on the sail's trailing edge – was flapping loose, having worn its way out of its sheath for a good couple of meters of the sail's height. Yes, this sail positively reeked of character. We knew we would need to be quite conservative with it, erring on the side of caution when it came to reefing, but nonetheless we were now equipped once again to battle the wind on more or less our own terms. Nice.

Later on the 25th I wrote again in the log:

25/09/10
Time not marked. Colin:

We have installed the spare genoa so are set for sailing properly when the wind comes.

696 DTD. 59° 51.264' N, 25° 12.053' W. COG 175°. SOG 4.3 kts. 1600 rpm.

Fuel 1.5 inch off bottom of window. Therefore 2 days in tank and 1 day of Gerries after that. So, if motor for the next day + half should have ~ 1 day of Gerries left for arrival or other manoeuvres.

God knows what *"other manoeuvres"* I had in mind! I suspect that is my euphemism for 'dealing with disaster'. Nevertheless, as stated, we only had 696 nautical miles left to cover. So, assuming an average speed of 6 knots, which is quite conservative, that's only just over 120 hours at sea, or five days. It was a massive relief to be looking ahead in terms of days and not weeks at this point. We were much closer to home than we were to our point of origin.

There finally seemed to be light at the end of the tunnel!

19 / The downhill slope

As we worked our way south, fighting down the digits labelled 'latitude' on the GPS, a breeze began to pick up. Thus, it was with much rejoicing that we were now able to deploy the sails once again, including the newest *re*-addition to our little family.

With the canvas hoisted and full of air we happily turned the engine off, thus arresting the consumption of our precious diesel supplies. We could almost smell home now. But that was a dangerous feeling. We had to work not to feel complacent, for as the ocean had already taught us, anything could (and worryingly, often actually did) happen. As it transpired we were saved from the clutches of temptation for, pretty soon after our sails started catching wind again that wind started getting ideas above its station and, we were once again taking a somewhat more than comfortable amount of 'moving sky'. We had in fact seen this forecast, but had hoped that perhaps this time we'd get lucky. Matt commented in the log (after quite naturally having a little worry about diesel, as we all were):

25/09/10
Time not marked. Matt:

Fuel motor 26th = one-half-day fuel. After that have ~2.5 days left total inc. Gerries. Therefore can motor last part above Ireland if need be.... Good westerlies predicted in-between.

Message from Nolwenn about 2nd low pressure system due on 27th@:

57° 24' – 56° 06' N, 22° 31' – 18° 43' W.

Has it moved? Will it miss us?

So no, it appeared the low pressure wasn't going to miss us. In fact it started cussing in our general direction a little earlier than the GRIB file suggested it might. What should have passed over on the twenty-seventh appeared to like getting to the party before anyone else had time to hit the nibbles. This one was tenacious! What's more, it started harassing us with southerlies, forcing us to compromise a little and head ESE instead of directly south, beating into the rising waves once again:

26/09/10
0029 UTC. Colin:

Diesel pump switched on at 0029. Day tank filled at 0043.

Fairly heavy weather on the bow: Gusting 30kts! However, it is predicted to pass, and we must go south!

Well…south-east in actuality.

We forged on without incident. In fact, this little breeze proved to be something of a blessing! Every cloud has a silver lining, and in this case the lining was that the winds galvanised us into another spate of proactivity. In lieu of our recent rope-related retardations we decided, under Matt's energetic encouragement, to back up positively every sail related rope or cord we could find on deck. This may have been overkill, but we really didn't fancy any more unscheduled snappages. We were damned if another poxy rope was going to stop us now! And how best to solve rope-related problems I hear you ask? Why, with more ropes of course! Fight rope with rope!

The cord which attached the mainsail clew to the end of the boom was augmented with another identical cord, tied even more conscientiously than the first. The mainsail's reefing lines were also doubled-up whenever a reef was put in, an additional length of rope being made accessible for lashing the reefing eyelet against the boom. The staysail got the same treatment, and we even discussed doubling up the genoa furling line. We were determined that there would be no more incidents. We were getting so close.

And so we were carried onwards and gradually eastwards. After a day or so we abandoned our push south and resumed a more easterly course. Simultaneously the winds lightened to a more comfortable 20 knots and even started coming at us from behind for a while. Unprecedented.

Things really did begin to settle down on the boat now, and in fact I think we may even have begun to genuinely enjoy ourselves a little. As ridiculous as it might seem we began to read more books on relevant subjects, such as sailing for example, and weather. Crazy I know, but hey, it was the party ship now.

From the log entries made around this section of the voyage it seems that we must've had a bit of surplus mental energy. A few new words crop up. Words like 'octas', which apparently is the unit which one traditionally uses to quantify cloud cover:

27/09/10
0214 UTC. Sam:

Have emailed Nolwenn with our position. Very nice outside. We are going straight towards the Moon. The wind has backed 030° to the east since 1333Z. Cloud cover has reduced significantly since 0632 (7 octas) through to 3 octas at 1757Z, and now 2 octas.

Nolwenn forecast SW 20kts becoming SE 40kts. We have seen the wind back to the SE but no strengthening. Diminishing cloud suggests we are clear of the fronts – if it builds and the wind strengthens we should be prepared for 40 kts. We should perhaps download the GRIB file…

Maybe it's just me, but I swear that towards the end of that last log entry Sam almost sounds like he knows a thing or two about meteorology. Incredible really, and it only took two storms and a fortnight at sea. Even I start throwing my weight around a little and brazenly venture to suggest that I'm more onto things than the boat's on-board systems. Ridiculous.

27/09/10
0530 UTC. Colin:

Wind 20-30 kts, but I'm suspicious of the anemometer now after its readings in the gale of Friday (which seemed far too high given the wind speeds). *However, the wind appears to have turned easterly, which was unexpected.*

We maintain a course of ~ 95°, but could use more south.

I have really damaged my chest muscle. Simple tasks are causing me a lot of pain now. This is a real disappointment, although glad it happened now and not one or two weeks ago!

Nolwenn texted to say he is really happy with our progress. I'm glad. It's great that he cares so much.

And so, other than a few signs of wear and tear beginning to make themselves felt on our bodies, things really went pretty smoothly from this point onwards. By way of example on the wear and tear front, as the log entry above describes, I had done myself a bit of an injury to my right pectoral muscle. This was the result of repeatedly working on the mainsail halyard winch, hoisting or tightening the main every time we needed to add or release a reef. It was agony, and although it really goes against the grain for me to back away from a physical challenge, I had to cut back on the amount of time I spent 'beside the mast' from this point on as I could feel the damage worsening every time I cranked the winch handle. But, although it's entirely possible that Matt and Sam might disagree, I really don't think that this minor reduction in manpower was much of a handicap at this stage. The weather began to give signs of levelling out for the foreseeable future, and spirits were high:

27/09/10
No time given. Matt:

57° 45.435' N, 19° 48.882' W.

Full sail. (Wind) *Not predicted* (to exceed) *25 kts all way to Oban on GRIB. Still large dying waves from low pressure system.*

RELIEF!

501nm DTD.

Of course you should never take these things for granted. Matt rightly says that the forecast was for pretty comfy sailing winds all the way into Scottish coastal waters, but the sea is a treacherous mistress. Things change rapidly. But, at least when spirits are high and emotional energy levels are charged up, good morale lets you take the most from what fate deals you. The next log entry illustrates both these points well, and in my opinion somewhat amusingly too, given the glow of barometric optimism which radiates from Matt's last entry above. Just a few short hours later Sam had this to say:

27/09/10
2128 UTC. Sam:

Best watch yet. Continuous Force 8 gusting Force 9 with 10 m of swell. We disappear into huge holes. Gambo doesn't care though, she just bobs along like a cork. We reefed the staysail and then the main; being occasionally plunged underwater. A squall hit us and the wind backed to the east slightly. The blowing rain is immense. It's impossible to look into the wind. The sea is like a war zone, even Bear Grylls would be scared.

I saw a dolphin jump out of the crest of a wave. It must be having fun!

Then, it's over. The wind drops to 20 kts and backs to the SSE. All sail goes up and we're making 6 kts towards Oban.

Proper heavy weather sailing. Awesome.

This is what it's all about!

I've often felt that when life is too easy it can be difficult to feel truly alive. In the land of paved surfaces, modern medication, labour saving devices, air conditioning, double glazing, tinted windows, health and safety and government benefits you actually have to work hard to find ways of putting yourself in danger. I think this takes something away from our souls, like a slow puncture in the great bouncy castle of well-being. Our bodies and minds evolved (or were built – it doesn't really matter), to survive! To survive is our sole purpose. To survive so as to procreate and therefore prevail. This is the primary objective of every organism ever studied by science. To stay alive! But what if that struggle, for which we are all so beautifully equipped, whether we know it or not, is taken away from us? Can we ever feel truly fulfilled? It's a difficult question to answer, both emotionally and ethically, given that I am painfully aware of my privileged life. But still, I personally find that it's hard to feel alive when there is no effort in living. Adversity breeds innovation, and innovation paves the way into the future. When we are forced to see things from a different perspective, for whatever reason, be it enduring a hurricane (in our case Hurricane Igor!), narrowly escaping some life threatening accident or simply living with new people whose way of existing differs from your own…. When we gain these new perspectives it opens our mind and brings us out of ourselves. 'Surviving' makes me feel worthwhile. There's no easy soundbite to sum this all up, and I hate the phrase 'what doesn't kill you makes you stronger', for there is a thing called 'post-traumatic stress', but hopefully you follow my drift. Maybe you even agree. That'd be nice.

20 / "Looking for feesh!"

Gradually we entered a sort of *'Twilight Zone'* of expectation. We had been at sea for quite a long time now and had been through a lot together. It had been tough going. There'd been quite a lot of Johnny Cash played on the stereo and we'd all had a painfully steep learning curve to climb. But we'd prevailed so far, and now there were tangibly fewer days ahead than there were behind. Our minds started to wander towards thoughts of home. This change in our attitudes was accompanied, and doubtless exacerbated, by a few events, the first of which was a message to us all from Tom, who had left us in Nuuk after (perhaps quite sensibly) deciding that he'd rather fly home in a day than endure the best part of three weeks' worth of good old-fashioned misery sailing back.

Sam logged Tom's message which read as follows:

28/09/10
0449 UTC. Sam:

Message from Tom Cowton:

Tom:

Hi guys. Nolwenn tells us you're still upright and floating towards UK. Good work! Hope you're not going too crazy out there. Tom.

Our reply:

We've lost many things, both physical and mental – still have loads of morale (Haribo jelly sweets) *left though! We can confirm that it is like trench warfare. ETA Friday afternoon. Love, the team.*

Tom:

Like the comparison – makes me happy to be back in a nice comfortable house. Glad the Haribo is holding out. Are you sick of noodles yet?

It was great to hear from Tom. His message was really supportive, and more than just a little insightful too: Yes. Yes we were sick of noodles! And we were also growing weary of the wind's indecisiveness:

28/09/10
0520 UTC. Sam:

Becalmed. Engine start. 57° 27.649' N, 018° 21.45' W. DTD 453nm.

28/09/10
0540 UTC. Sam:

Wind picked up.: Engine stopped.

Wind 20/21 kts. SSW. Weather seems as forecast. Barometric pressure = 993 mb. Forecast pressure = 997 mb.

The conditions just couldn't seem to make their minds up. Our sails seemed to be up and down like Captain Kirk in a crisis, prompting incessant bouts of diesel consumption as we would engage the engine so as to maintain our forward momentum. It was far too late in the game to be waiting around for the wind, and while we still had fuel left we were going to use it. So long as we kept twenty-four hours' worth for negotiating the inland waterways we would be OK.

Another strange thing happened on the same night that Tom messaged us; another sign that we were nearing land: for the first time since crossing the continental shelf off Greenland's west coast we saw another vessel! I recorded the incident in the log:

28/09/10
0643 UTC. Colin:

This morning marks 14 days at sea! A long time, but we are getting close now. We even spoke to another boat last night! A fishing boat which seemed intent on hitting us!

It was a surreal encounter which evolved over the course of perhaps twenty minutes and during Matt's watch. It was the middle of the night and we were sailing in light winds. Both Sam and I were in our beds when Matt quietly crept into the main cabin and woke me with a gentle prod: *"Colin. Colin! There's another ship out there mate!"*

I had been sleeping quite deeply this time, and it took me a few moments to really switch on to what matt was saying.

"Mmm...." Yawn. *"What? Where?"* I replied.

"Off to our port side, ahead. It looks quite far away, but I don't know what kind of ship it is."

I was quickly waking up now and beginning to feel genuinely excited that we had finally seen another ship. I climbed out of bed and, in my underwear alone, clambered up the companionway. My blue Scottish complexion must have been positively luminous in the dazzling moonlight as I stuck my head around the dodger to look for this other vessel.

And there it was, just as Matt had described, a little to port of our bow but still some distance away. It was small, and was broadcasting a combination of coloured lights from its stocky masthead to distinguish itself as a fishing boat. It was perhaps a mile or so ahead of us, and, judging by the red navigation light I could see roughly level with its superstructure, it was moving from right to left, crossing our bow. So, it seemed there was no need to worry. We wouldn't hit it on this course. No dramas. In fact, this was good news. It showed that we were once again approaching the shallower waters of Europe's continental shelf; shallower seas with more life, and therefore more for fishing boats such as this to hunt. Other ships would soon become a more frequent. This ostensibly ought to be a bad thing, given that they pose a potential collision risk, but after weeks of solitude we were all becoming a little lonely, and welcomed the sight of man-made objects again. The lights of this little fishing trawler were like a cityscape in microcosm. They were warming to see.

So, satisfied that there was no danger and buoyed up by having seen proof that we were indeed nearing home, I returned to my sleeping bag and quite quickly fell asleep again. However, in what seemed like only an instant but after what was probably more like fifteen minutes Matt woke me up again, this time sounding a little more urgent.

"Colin. Colin! That boat's still there, and it's pretty close now!"

"How close?" I moaned.

"I'm not certain, but pretty close. Maybe twenty metres?"

This sounded unlikely to me, given that we were still very much in the middle of a big, empty ocean, but I got up to check anyway.

Once again I abstained from any additional clothing and climbed outside to look. And there the bloody thing was, about twenty metres away from us off our port quarter! I could almost make out the silhouettes of people through the glowing portals of light in the cockpit area. It was more or less behind us, so there still didn't appear to be any risk of a collision, but still….

What the hell was it doing so close?

I immediately jumped downstairs, picked up the VHF handset and, after only a moment's pause, which I spent contemplating how to address them, I pressed transmit on channel sixteen and hailed them.

"Fishing vessel. Fishing vessel. This is yacht Gambo, yacht Gambo on channel sixteen, over."

I decided that, given we were in the middle of nowhere, and there could surely be little ambiguity about who was calling whom from where, 'fishing vessel' should suffice as a term of address.

After perhaps ten seconds of silence I repeated my hail, and this time there was a reply. It was difficult to make out as whoever was on the other end plainly didn't use English as their first language, but they had certainly heard us and were inviting me to say whatever it was I wanted to say to them. So, I posed my question as diplomatically as I could.

"Fishing vessel. Fishing vessel. This is yacht Gambo. Please explain your intentions. You are very close to us."

A few long moments elapsed before the radio crackled into life again. Perhaps they were working out what to say, or perhaps they were wondering how to say it. It doesn't matter, because the reply was hilarious.

"Uhhhhh, Gumbo, Gumbo. No problem. No problem. We look for the feesh, we look for the feesh, we look for the feesh. Over"

Brilliant! This much was surely obvious. Vessels do carry different configurations of night lights for a reason after all. I knew they were fishing already. Still, at least they hadn't said that they wanted to board us and attack us in our beds! So, I replied, diplomatically once again.

"Fishing vessel, this is Gambo. Yes, OK. Could you please just do it a little further away from us? Over."

"OK. No problem. We look for feesh."

There really wasn't much else to say by this point, so I rounded things off with.

"Thank you. Yacht Gambo listening on channel sixteen. Out."

Surreal! I still don't understand why they came so close to us. For a while we speculated wildly. Perhaps they were smugglers? Maybe they were due to rendezvous with someone to exchange drugs, or weapons or something and this was a case of mistaken identity? That'd be interesting! It was all probably just wild fancy in all likelihood, the chances being that they were simply bored fishermen who fancied having a closer look at us. Perhaps they hadn't seen anyone else in a while either! Whatever the reasoning, it was a memorable encounter.

"No problem. We look for the feesh!"

Classic.

21 / And the wind whispers—

"Maybe?"

Later on the twenty-eighth I looked at the A4 print-outs (which we had in lieu of actual nautical charts) and studiously plotted waypoints into our GPS, to define our route through the islands and waterways of Scotland's west coast. Very soon we would be in amongst the land, and like a child at Christmas who makes sure that milk, mince pies and a carrot are left out for Santa and his reindeer, I wanted to be ready.

Nolwenn had suggested a route to me before we left Nuuk, but I'm afraid his ideas fell on deaf ears; because on this matter of all matters, I was going to please myself. I had grown up near Oban, and the landscape of fjords and islands of Argyll and Bute lay like a map of my heart on the GPS pixelated display. This approach, given the places we would be sailing through and past, had almost ritualistic significance to me. So although Nolwenn quite sensibly suggested a track which would carry us south of the Isle of Mull and thus away from the busier shipping lanes, I wanted to take a more 'scenic' route. This was my old stomping ground, after all! So, I opted for the Sound of Mull, amongst all the ferries and other yachties. I wanted to see them all, and I wanted them all to see us. We'd just sailed in from Greenland! Didn't they know?

It was going to be incredible, because if I had been given free rein to pick any port into which to sail I would have picked Oban myself in a flash. It couldn't have been set up as any more of a poignantly emotional homecoming for me if that had been the sole purpose of the voyage (which of course was very far indeed from being the case). Oban just happened to be where Alun had managed to secure a mooring, through a friend of a friend.

So, I felt singularly satisfied with the cards serendipity had dealt me.

The waypoints I chose to plot into the GPS were as follows:

"1: 'Sea of the Hebrides entrance': 56° 30.72' N, 007° 35.38' W.
"2: 'Hawes Bank': 56° 45.61' N, 006° 31.94' W.
"3: 'Fishing Bank': 56° 40.29' N, 006° 11.01' W.
"4: 'Rubha Nan Gall (safe water)': 56° 38.86' N, 006° 03.34' W.
"5: 'N. Salen Bay': 56° 32.99' N, 005° 55.98' W.
"6: 'Ardtornish Bay Beacon': 56° 31.08' N, 005° 45.57' W.
"7: 'Duart Point': 56° 27.49' N, 005° 37.25' W.
"On to 'Sgeir Dhonn' (existing wpt)
"'Lady Rock'."

But of course this was all still a few days away yet! I was just getting excited. We were still very much at sea. But, the landscape of that sea was changing. As we moved eastwards over Rockall Bank, basking under blue skies and cheerful September sunshine, and finally running with some light westerly winds, the texture of the ocean took on a face we had never seen. The whole might of the North Atlantic was washing up against the continental shelf, and, where over a kilometre's depth of water found itself funnelled into 'shallows' of only around 200 metres, great muscular swells were born. We were riding forward on colossal waves, each one perhaps 10 meters or so high, but five or six times as long so that they didn't break, or even peak, but rather just surged past us like great rounded, mobile hills. As a wave crept up behind us we would surf down it at breakneck speeds. It was exhilarating, and the water boiled around our bow as Gambo cut downhill into the water. On one wave I watched our GPS display and saw the speed breach 9 knots. For a boat built like Gambo, that's fast.

As each wave rolled beneath, beyond, and away from us, we would languish in the trough of its wake, feeling almost like we were being sucked backwards. But, only a few seconds later, the next swell would come surging in and again we would be propelled forwards at warp speed, loving every second. It was brilliant. We all spent a while up front, standing on the bowsprit and enjoying the feeling of being right out there on the cutting edge. The water parted for our bow like air on a swinging sword. We were literally surfing. It felt refreshing to be moving so quickly and efficiently. We were now being aided on our way, by the very self-same sea which had seemingly been trying so hard to break us for weeks. It felt almost like we had passed some kind of test. We had graduated, and this was the ocean's way of saying *"Good on y' mate!"*

Talking of refreshing, something else unusual happened on this day. It wasn't anything to do with the sea state, or the weather, nor were we visited by any unusual animals – although you could say that it was in a sense wildlife related. No, nothing untoward occurred, but it was a surprising turn of events nonetheless: Sam decided to have a wash!

Now, I don't mean to suggest that Sam is known for poor hygiene. Quite the opposite in fact, for Sam has some of the best 'personal admin' of anyone I know. Poor hygiene and soap aversion is more my territory, to be quite frank. But, still, I was quite surprised at this decision to suddenly get clean. I for one like to save that very particular joy for arrival at our final destination. (Yes, I know what you're thinking: you're thinking that by making this admission I'm giving away the fact that I'd now allowed two weeks to pass without washing. And yes, this is true, you're right. But remember that it was just the three of us on Gambo. Who was there to impress?)

One of the things learned quite quickly when sailing long distances is that you just don't have privacy. Matters quite taboo or outright unacceptable back home really need to be overlooked on a small boat. For example, if the toilet is either broken or inaccessible (if say, for example, it is so full of climbing gear that you can't get in) then you just need to 'go' over the side of the boat. And, if someone happens to be on the helm at that time . . . then they are just going to have to avert their gaze. But of course, it'd be rude not to pass the time of day, so you really have to strike up a conversation in these situations. Therefore one becomes accustomed to holding a good old chinwag with mates who happen to be laying a log overboard as you shoot the crows. You get used to it, surprisingly quickly in fact. (It's also surprising the variety of seemingly un-appetising things which albatross – those graceful ship shepherds of the southern swells – will eat.)

My point is that body odour really isn't that big a deal, especially if you all stink just as much as each other and there are no girls around. This factor would of course make a massive difference. But for now let's assume an all-male environment. If you think you might be uncomfortable without having a wash for so long, any truth in this is eclipsed by the sheer ecstasy of arriving safely at your port of destination and having that first hot shower. You emerge a new person altogether, cleansed physically, mentally and spiritually. Yes, that first shower brings you face to face with the holy trinity of personal hygiene. It's divine, metaphorically speaking. And now Sam was about to ruin this for himself. And with only a few days left to go? I just couldn't understand what he was thinking! What's more, he was going to do it on deck, with cold salt water! And, he wanted me to take a picture! This just couldn't have been weirder. But, I consented to capture the moment on film, although I made no secret about the fact that I disapproved of this folly in principle. Not of the picture taking I might add, just of the washing.

Sam did keep his shorts on. There was no nude photo shoot on the poop deck. Yes, the picture does have a certain *Men's Health* magazine quality to it, but that's OK. We were all friends on board the good ship Gambo.

Actually, it was all quite innocent. I'm still unsure to this day why Sam didn't just tell me this at the time, but I later learned that the good Mr Doyle has a collection of similar photos – pictures of himself having a wash – taken around the world in strange or surreal places. For example, he has a photo of himself taking a bath in a water drum whilst on the West Antarctic Ice Sheet, where he once stayed for several months working for the British Antarctic Survey. Now that I know this I think it's great banter!

There are few things as satisfying as performing mundane rituals in ridiculous places. For some it's 'extreme ironing', but for Sam it's washing in ridiculous places. And yeah, sorry mate – I hope he doesn't mind me airing this particular little idiosyncrasy! It was just too memorable an event not to share.

Would it make it any better if I said that I have a collection of pictures of myself naked on mountaintops around the world? (*Ahem!* Anyway. Moving on, swiftly. . . .)

Now very much into UK waters, we started thinking a little more practically about where we were going and when we expected to arrive. How much fuel was left? How much would we need? What would we do in the event of another emergency? Because of my time spent with the RNLI, I had a pretty good knowledge of how coastal shipping and leisure traffic worked, and so I did what I felt I ought to do, that being to call the coastguard and log our whereabouts and intentions.

As it happened, I had actually done some research before we'd left Nuuk. Perhaps not into immediate matters such as tide times and strengths, but instead into which numbers we could hope to call from the satellite phone in the event of an emergency. This was perhaps a little melodramatic of me, perhaps even verging on the pessimistic and fatalistic, but here I was looking up those very numbers I'd diligently googled almost a month before, so it had been worthwhile after all. Now, I flicked to the back page of my diary and fingered down to the number for Stornoway Coastguard. I'd also jotted down the number for the Canadian Coastguard HQ, the US Coastguard equivalent, Reykjavik Coastguard and Scotland's other (and now only) west coast coastguard station, Clyde Coastguard. I can't tell you how good it felt to make a phone call again, tapping in the numbers on the brick-sized satellite phone handset and listening to the dial tone. In just a few seconds I was connected, and a Hebridean accent was bubbling down the line to me. It was like music to my ears.

All I did was relay our position, state our port of origin and state our intentions, but by doing so I ensured that at least we were now 'on the grid'.

I would contact them again by VHF as we moved past the outer Hebrides, hopefully in a day or so's time, and if I failed to do so they would know something had gone wrong. I would then contact Clyde Coastguard as we moved into their area of responsibility and make the same arrangements. From this point onwards, in theory anyway, we wouldn't be alone. This, at least for me, was a massive relief. After two weeks of lonely exposure in the Atlantic it felt good to be on someone's radar. I think we all appreciated it on some level, even if it did perhaps sting some of our pride to even look in the direction of the emergency services, no matter how briefly.

Things were good. We were all happy and well, and now at least one of us was clean too. That night, Sam wrote a particularly long and particularly involved log entry. I think it's beautiful, and it nicely sums up how we felt as this day came to a close.

29/09/10

0700 UTC (approximately). Sam:

A school of dolphins rides our bow wave as the sun rises in the east.

A perfect day. The most rested I have been in several years, followed by a perfect night. Steady, 15 – 20 kt southerlies allowed a heavy sleep. The sun rose into a clear sky. The deck was scrubbed, the galley tidied and the whole boat generally spring cleaned. The warmth of the sun was sufficient for sun bathing, so I washed on the deck in salt water and followed with a welcome rinse with fresh water. Very refreshing. Soon after Colin made broth and flat breads with the more exotic of our leftover vegetables.

The steady southerlies stayed with us through until evening and sunset, although Nolwenn and the weather forecast predict otherwise for tomorrow.

As the sun sets more watches begin: Colin's first, then Matt, then mine. After a cup of green tea, more broth and a bowl of sliced oranges. I take a peek outside, windward into the warm breeze, accompanied only by a book on 'Victorian ethics'.

Darkness reveals a sky full of stars, some twinkling red and green, others shooting across the sky. Venus shines brightly off the starboard bow; the plough visible off to port. I'm reminded that there are other sources of light other than 'our' Sun. The stars for instance, but also the blue glow of phosphorescence emitted by the plankton being disturbed by the boat.

A red light emerges off the port bow. I tune the squelch on the radio, check the volume and peer through the binoculars. I wake Colin and Matt. The sighting of our second ship since Nuuk is an event! But Colin quickly realises that the red light, now becoming disc-shaped, is the Moon rising above the horizon. I wonder if the light is red due to its greater travel time through the troposphere as a result of its low elevation. . . . Sometimes perhaps it's best not to explain phenomena with science though, or even think for that matter. Somehow its better just to enjoy. Either way I'm not about to make a radio call to the Moon!

As I write this a school of dolphins ride our bow wave. Their bodies are lit up by the phosphorescence they disturb to the effect that their every movement is visible. They seem incredibly efficient, occasionally breaking the surface and becoming dark silhouettes. We are cruising at 6 knots. The wind is building. It couldn't be more perfect, but now it's time to sleep. N56° 16.98' W010° 59.52'. COG 088°. SOG 6.0 kts. DTD 196 NM. Next WP: 'HEBRIDES' Dist 114 NM Bearing 082°. ETA Oban: Friday morning."

Toilet hatch leaking quite a lot – suspect it always has and always will. . . .

22 / Land ho!

As the sun slowly edged up to and inexorably beyond the horizon on the morning of the 30th September 2010, we all awoke with baited breath, knowing that at some point in the next twenty-four hours, visibility permitting, we *should* see the grey silhouette of land for the first time in a long time. We knew it ought to happen, and had plotted on the chart where we expected it to appear, but no-one dared speak of it too much. Perhaps we'd jinx it! Trying to stay rooted firmly in the moment and perhaps not too predisposed towards the future, Sam sent a nicely rounded email to Nolwenn updating him of our position, our intentions for the approach to Oban, and other significant events of the last few days.

30/09/10
0502 UTC. Sam emails Nolwenn:

N56° 21.9' W009° 32.7', COG 081, SOG 7.0 kts, wind 25 – 30 kts / 155°, DTD 148 NM, ETA Oban 1400Z, Friday morning. Routing through Sound of Mull. All well. Sunbathing yesterday plus annual wash.

With this done, Sam came off watch and roused me to take his place. I groggily emerged into the creamy early morning light and was met by a very welcome sight:

30/09/10
0755 UTC. Colin log:

Nice morning. Lots of sea birds including the first gannets! No sighting of land yet though.

Gannets! Fantastic!

Gannets, for those who don't know, are a large and streamlined diving sea bird found in various places around the world but with several large colonies in Scotland. One of these colonies lies on the Bass Rock in the Firth of Forth near Edinburgh, one on Ailsa Craig to the south east of the island of Arran, and a third is situated on the Island of St Kilda, about 40 statute miles west of Scotland's Outer Hebrides. I had very much hoped to at least see if not actually visit St Kilda as we neared the Hebrides. But, unfortunately, our assorted calamities and frustrations of the past week had caused that particular ambition to slide far down our list of priorities, and now we just wanted to get home. So, we didn't divert to St Kilda. But, some of St Kilda's citizens had apparently diverted to us! And they were a joy to see.

I've always loved gannets. They were one of the first birds that I remember learning to recognise as a child. As I've mentioned, my family was based in Argyll and Bute in the west of Scotland when I was growing up. We lived in Lochgilphead, a little over an hour's drive south from Oban. My father travelled a lot to the Hebrides for both work and pleasure, and occasionally my mother and I would come along with him. I remember standing on viewing decks of the Caledonian MacBrayne ferries with my dad and watching the gannets fly alongside in groups, their beady eyes focussed intently on the waters below. Every so often they would spot fish and, almost in unison, groups of birds would fold their wings impossibly far backwards and drop from the sky, like Second World War German Stuka bombers, shooting into the sea like feathered harpoons. They are so streamlined as they drop, and thus enter the water at such a speed, that they hardly make a splash. Sometimes I'd see them from our canoes or from our rowing boat when my dad took me out 'fishing' for wild oysters or the like. He'd pretend to shoot the gannets out of the sky with an invisible machine gun and very often the birds would oblige by crashing into the water right on cue. To me, as a kid, it was hilarious, and I find myself doing it still to this day whenever I see gannets flying low over the water. My invisible Kalashnikov comes out and the birds spiral out of the sky riddled with imaginary bullets.

These birds represented a line on the map for me. That was it; as far as I was concerned we were home. Everything from this point on was just sightseeing, and bloody good sightseeing it was too. Then, a few hours after I'd emptied my imaginary ammo hopper into our welcome feathered visitors, a new and very welcome shape appeared on the north-eastern horizon.

30/09/10
1400 UTC. Sam:

Land sighted! Barra Head. Colin raises the Scottish and Greenlandic flags. Upside down at first, but soon corrected. N56° 32.3' W007° 46.3'. COG 072°, SOG 6.7 kts. DTD 88.1 NM.

Finally, we could see it. Scotland. Our first sighting of land since September 16th, a clear two weeks previously. Admittedly it was little more than a dim and distant triangular patch of grey on the horizon, but land it was, of that we were in no doubt. Sometimes it can be difficult to know for sure, especially given that spending enough time at sea can make folk somewhat prone to seeing what they want to see on the horizon. But we had technology on our side. On went the radar, and there she was. Our little grey triangular blur had mass and form and everything. She sent back a bona-fide radar return like no cloud ever could. It was Barra Head, the southern headland of the island of Berneray and the southernmost point of the Outer Hebrides island chain.

Waypoint #1: *'Sea of the Hebrides entrance': 56° 30.72' N, 007° 35.38' W.*

Check!

I was home at last, and being something of a traditionalist (and an impatient one at that) I finally did what I had been waiting for two weeks to do. As Sam mentions in his log entry, I ran a couple of flags up the mast. One was the flag of Greenland, to indicate to any who cared to look where we had come from. The other was the St Andrew's cross, or the Saltire: Scotland's flag. Neither were 'proper' flags, by which I mean that neither had been purchased in a shop, but whatever they lacked in mass-produced quality they more than made up for in character, which to me was far more important. Both had been hand-stitched, the Greenlandic flag by someone else a couple of season's previously and the Saltire by me. I'd been working on it for a few days now, using a rectangular piece of a blue rag which I'd found lying around and two white crosses which, quite appropriately I think, I had carefully cut from the remains of our former headsail. This raising of flags is one of my favourite rituals when sailing. Partly I just love the tradition of it – I've always loved flags for some reason – but also it heralds your approach either to somewhere new and exciting or simply to somewhere safe and familiar. It means *journey's end.* Yes, there was still a muscular distance yet to cover, but those miles were all in Scottish waters now, and that felt good.

Conditions were calm now that we were amongst relatively sheltered waters. The big swells of two days previous were long gone and, despite a stiff wind which was hitting us from the north-east, the sea state was pretty reasonable. So, we carried on about our business as usual. For all intents and purposes everything aboard ship was the same as it had been for days and weeks now. But inside, in our hearts and minds, everything had changed. As the depth of the water beneath us dwindled our fears waned also.

A few hours passed and we made our way that little farther east-north-east until a second landmass appeared off to starboard of our bow. At first it wasn't as much a landmass, but more a single, lonely-looking white sphere. It looked pretty odd at first, and even with the annotations on our charts/print-outs to go by it could have confused many a stranger to the area. But I'd spent many months of my life on these islands and I knew exactly what we were looking at. With a surge of nostalgic excitement I realised that we were seeing 'the golf ball': the large white, spherical radar dome which sits incongruously atop Carnan Mor; the highest point on the island of Tiree.

This was the first land-based structure we had seen since leaving Nuuk behind us on day one of the voyage, and although perhaps not the most attractive of beacons it nonetheless represented our first visual contact with 'real-world' civilisation in two-and-a-half weeks. This was a structure I knew, and had indeed stood beside in the past. I had spent rather a lot of time on the island of Tiree. My father had been a staunch nationalist and a life-long proponent of Scottish language (Scot's Gaelic), Gaelic culture and vernacular skills and crafts. So, working as he did in regional planning and later as a historic buildings inspector, he became very closely involved with the protection and maintenance of Scotland's numerous historic thatched buildings. He in fact learned to thatch in his twenties and kept the craft alive throughout his life, going on to become something of a world authority on the architecture and regional variation of thatched buildings throughout Scotland. Anyway, through my dad, both my younger brother Peter and I learned to thatch and at one stage or another have both worked as self-employed thatchers across Scotland. One stronghold of thatched roofs is the island of Tiree. So, it's somewhere we've spent a lot of time. It's a beautiful place, but as either Sam, Matt or I can tell you after our sojourn on the North Atlantic, somctimes it's possible to spend a little too long in isolation. My dad always described his relationship with Tiree as being one of *"love and hate"*, and although I reckon he could have said that about more than one relationship in his life, this particular brand of affection which he held for Tiree was one with which I think both Peter and I can empathise. Anyway, I knew Tiree very well and, seeing it appear on the horizon, crowned by that golf ball (from which on a clear day you can see most of Scotland's Hebridean Isles), I felt my pulse quicken.

Slowly the golf ball rose from the waves, buoyed up by an ever growing sliver of land which, as it emerged from behind the visible horizon, gradually extended both to the north and to the south before ending abruptly, only to be joined by a second low-lying greyish line to our left (north east): the island of Coll, Tiree's closest but in many ways totally dissimilar neighbour. Whereas Tiree is an island of sandy beaches, low-lying bogs and fertile fields, Coll is a rock: a beautiful protrusion of rolling gneiss coated in the thinnest layer of peat and heather. It is beautiful in a wholly different way to its partner. Tiree is a land for beach walkers and wind surfers. Coll reeks of wilderness. From where we stood on that day however, both were like blazing beacons of homely salvation. We turned to port, assuming more of a north-easterly course and drawing level with, and parallel to, the island of Coll.

Waypoint #2: *'Hawes Bank': 56° 45.61' N, 006° 31.94' W.*

Check!

Soon daylight began to fade and the wind began to pick up. The greyish silhouettes of land which were now standing off all about us (Coll and Tiree to starboard and Mingulay, Vatersay, Barra and South Uist all to our port side) were slowly but surely replaced by thin, sparsely populated strips of light. Some of the uninhabited islands of course vanished altogether, namely Mingulay and her outlying islets, their rocky outlines gradually merging with the ever-pervasive darkness. Night was falling, and as living room lights were switched on around the country, folks would be curling up in their warm, stable, centrally heated and pleasantly furnished homes. We all felt a pang of envy at that point, I think. Here we were; still at sea; still cold; still getting wet; still being battered by the weather. We were still an island unto ourselves. But no matter, for with only 35 nautical miles left to cover, by the time the sun rose again on the following day we should be within sight of our goal: Oban.

We switched on our night-time navigation lights, put on another layer or two of clothing, and hunkered down to visually scan the horizon. We were tired, and our eyes were red from seventeen days of wind and salt spray. But, for the first time now there were actually things on the horizon to look at, and our eyes found renewed vigour in this target-rich environment. We were searching for the all-important navigational lights and signals which would soon begin to appear; blinking beacons of coded information which, when cross-referenced with our charts, would tell us both of our location and also that of things best avoided; things such as rocks, islands and headlands; all characters with which you don't really want to get too personal, even in a robust steel vessel like Gambo.

Soon enough, a single bright and flashing light emerged from the gloom off our bow, waxing from the black horizon as if carried from the seas atop a great tusked kraken. It was the stalwart Ardnamurchan Lighthouse, teetering proudly on the westernmost headland of the Ardnamurchan Peninsula. This flashing light was our first glimpse of the Scottish mainland. And, not only that; it was our guide. Intermittently carving the night like a great swinging sword of light this brazen beacon denotes the northern side of the entrance into the Sound of Mull, the long waterway through which we intended to pass on our approach to Oban. I was very pleased to see it.

The long-standing relationship between seafarers and lighthouses is one which is well known in every maritime nation's culture and folklore, even in communities not directly located on the sea. Lighthouses are symbols of hope and salvation; quite literally a 'guiding light' to those in need of direction or aid on the seas. Since their advent in the seventeenth century, and subsequently becoming more widespread in the nineteenth century, they have saved the lives of countless sailors. They forever changed both the world's coastal landscape and man's relationship with the sea in a manner every bit as revolutionary as the introduction of steam propulsion. Viewed from land, they are a wonder. Viewed on the other hand from the sea, they are a godsend.

Even in daylight lighthouses often make an impression. Their tall, impossibly slender trunks jut defiantly from the rock – often in the most unforgiving and seemingly inaccessible places – like brave trees on lifeless shores. An onlooker might justly wonder at how men were able to erect such unlikely structures in exposed locations such as those which often demand a lighthouse's presence. And yet built they were, and stand they do: testament both to human ingenuity and to our indefatigable ties to the sea.

It was, in fact, in nineteenth century Scotland where many of the earliest and most important advances in lighthouse structural engineering were made; the Edinburgh-based Stevenson family (a family also famous for the Scottish writer, Robert Louis Stevenson) being responsible for the design and construction of numerous impossibly tall, impossibly exposed, and yet extraordinarily robust lighthouses around the British Isles. Most of these erections still stand today and continue to guide shipping as they always have. Amongst them are the Bell Rock lighthouse which is situated off Arbroath on the Firth of Tay; Skerryvore Lighthouse to the south of the island of Tiree; the May Isle lighthouse in the Firth of Forth; the Eddystone lighthouse which is located in The English Channel, and the Ardnamurchan lighthouse, towards which we on Gambo now found ourselves sailing.

All of the lighthouses which I mention above are striking landmarks which anyone can appreciate, regardless of stock. That said, however, I don't believe that there can be any substitute for the gratitude and appreciation which is felt for a lighthouse when you are yourself at the helm of a ship, battling the seas as the light fades around you. Your visual references slip away into the flat void of night and a very natural, very real, and very justifiable fear comes over you: The fear of the unknown. It's true that in the modern age most vessels have an array of electronic devices on-side to help even the most spatially challenged amongst us navigate safely between landmasses, but when you first catch a glimpse of a lighthouse and read its identity from that particular sequence of flashes . . . that's when your confidence returns. Like a loved one waving from a hilltop it calls to you and ushers you through, and in the pit of your stomach you wonder at those who sailed these waters in bygone darker days; days before the age of GPS and before we saw fit to erect lighthouses such as we enjoy today. As the Sun betrayed the sailors of old; as the clouds gathered round to block the stars and the winds whipped about their wooden worlds; what hope was left to those who would navigate close to land? A life at sea was as good as an early death sentence to many. I for one will never take lighthouses for granted.

In fact, equipped as we now are with a staggeringly extensive and comprehensive suite of navigational markers around the country (and the world for that matter) it can actually be *easier* to navigate coastal waters at night, especially when coming into a new harbour. During daylight hours coastlines can appear two-dimensional, and it can be difficult to discern headlands, inlets and relevant topography until you are uncomfortably close. At night however, lights of specific colours, and set with specific coded flash sequences, broadcast information at you from a distance of miles, usually leaving you in no doubt of where you should be heading. Of course where there are too many lights – such as when approaching a city – things can get confusing. For example, I once very nearly sailed clean into the side of a cargo ship when entering the harbour of Portland, Maine, as I had been aiming for 'the dark gap in amongst all the lit obstacles'. It turned out that this dark gap was the un-lit side of a ship's hull! Fortunately we realised our mistake about 30 meters short of an almighty bang when it became apparent that the 'gap' was moving from right to left! So, yes, lights can be confusing. But generally you know where you stand with lights, and I like that.

As fate would have it my affection for night navigation was something of a blessing. As has happened to me all too often, nightfall manifestly appeared to be coinciding rather perfectly with our arrival at the entrance to the Sound of Mull. The wind had also swung around to the east and as we made our turn into the channel we found it blowing directly onto our bow. And it was no simple breeze, but rather a somewhat impudent 30 knot howler. Typical! So, with our patience for the wind now almost complctcly exhausted we dropped all sail and switched on the engine. Our brains and bodies had been drained, but our diesel reserves had not! So, where another crew might happily have resigned themselves to tacking expertly back and forth along the channel; weaving a merry zigzag all the way into Oban, we gave an almighty bird to that idea and stowed the canvas for what we were fairly certain would be the last time and pitched ourselves into the ill-tempered elements. As it happened, the tide was in our favour, at least for the time being, and this gave us an extra boost as we thrashed ourselves a foamy groove into the Sound of Mull. Unfortunately however, this combination of wind against tide has a tendency to produce particularly unpleasant waves. With wind and sea spray whipping us from the bow I soon found myself unable to look directly ahead of the boat without protection, and so it was through a diving mask that I found myself searching the darkness for navigation lights, wiping the Perspex every few seconds to clear the spray which assailed us mercilessly as we forged our way deeper into this fjord-like landscape.

Waypoint #3: *'Fishing Bank': 56° 40.29' N, 006° 11.01' W.*

Check!

One by one the red and green lights of this busy shipping lane slipped by us. Every so often a cardinal marker would appear, instructing us with its specific sequence of flashing white lights whether we ought to pass north, south, east or west of it. The wind persisted for a time and I found it exhausting to spend too long on deck, so I decided that perhaps it was time to use some more of our electronic aids. So, on went the radar. Now we were really cooking on gas. There, in resplendent shades of fuzzy green, was a somewhat abstract representation of the hills of Morvern on port and the Isle of Mull to starboard. And right in the centre – encircled by a great radio arc which swept the display, erasing and subsequently re-drawing the world around us every couple of seconds – was us; exactly where we should be.

Conditions-wise, these were very close to being about as bad as we could ever have hoped to encounter for our approach to Oban: pitch blackness, a howling headwind, driving rain and poor visibility. It was awful. But despite it all I was having the time of my life. I was here in the waters of my childhood and amongst the hills of my homeland. The darkness wasn't so much frightening any more, but comforting. I could make out the mountains on either side of us, looming out of the night not like monsters, but like watchful parents. I felt truly wrapped up in the landscape of my home, and with it being night-time I could almost believe that I was here alone and that it all belonged to me. My watch had actually come to an end several hours before, but I was enjoying myself so much I told the other guys just to take a break and get the sleep. There really was nowhere else I'd rather be than up and about, navigating us into Oban. Matt gave a grateful thumbs up and seized the extra rack time. When it came Sam's turn he seemed hesitant, probably unsure of what the polite thing to do was, and eventually decided to stay up and set about making hot drinks (winner!) and cleaning the cabin. So, as we neared journey's end and the prospect of shore-based facilities, Gambo's on-board toilet had its first scrub in I don't know how long. This was perhaps unnecessary at this stage in the game, but both Matt and I appreciated the gesture nonetheless.

Very soon, just shy of the town of Tobermory (situated on the Isle of Mull), we made a slight turn to starboard, following the channel as it doglegged it's way ever farther eastwards toward Oban:

Waypoint #4: *'Rubha Nan Gall (safe water)': 56° 38.86' N, 006° 03.34' W.*

Check!

With this turn we found a little shelter. The wind lessened and the waves became markedly more subdued. I was able to remove my diving mask and for the first time stand on deck and look around un-assailed by airborne seawater. Geoffrey was steering, prompted by me for occasional corrections to port or starboard by simple presses of a button on the control panel. It was blissful.

I stood by the companionway enjoying a hot cup of tea, scanning the horizon, occasionally checking the chart to see which light should appear next, and nipping back up on deck to ensure that reality corresponded to our instruments, and vice-versa. And this is about as tough as it got from this point onwards. On a couple of occasions another ship appeared either ahead or astern of us, but this was more an occasion for excitement than anything else. Somewhere just to the west of our fifth waypoint a large vessel began to gain on us, and as it drew closer I realised that it was one of the iconic Cal Mac ferries whose black and white hulls and Red-painted funnels were such a symbol of my early years in Argyll and Bute. The ferry had doubtless seen us, and there is ample room in the Sound of Mull for many large ships to slip past one another. There really was no danger of a collision, but I simply couldn't help myself: I nipped down the companionway and after a moment's nervous pause I picked up the VHF handset and hailed them:

"Caledonian MacBrayne ferry, Caledonian MacBrayne ferry on an eastbound course west of Salen Bay, this is yacht Gambo, yacht Gambo on channel sixteen, over."

A few seconds passed in silence, but then sure enough a reply came through:

"Yacht Gambo, yacht Gambo, this is MV Lord of the Isles. Go ahead, over."

So, I invited the *Lord of the Isles* to join me on channel ten, off the public ship-to-ship working channel, before just ensuring that they were aware of our presence and our intentions. I really needn't have bothered, as these ferries transit the Sound of Mull literally every day, and yachts are a common (and doubtless irritating and amusing) obstacle. But, I'll be honest; I just wanted to say a hello. The Lord of the Isles' radio operator was probably shaking his head as he hung up the VHF, lamenting my paranoia, but little could he know just how excited I was to see him, and how it made my insides glow to reach out and make contact with one of the boats on which so many of my best childhood memories sail.

In the event, my exchange with the Lord of the Isles was a very pleasant one, and if there was any impatience on the other end I certainly didn't pick up on it. The Cal Mac ferry service is a very real and, in my opinion, absolutely crucial pillar of Scotland's west coast community. This sentiment of mine was reinforced two years after this simple radio exchange when in 2012 I overheard a very unfortunate incident play out over the radio, also in the Oban area and also involving the Lord of the Isles ferry. I will recount it here as I think it highlights the importance, and also the enduring strength, of the camaraderie and mutual sense of responsibility which exists between all those who travel by sea, be they yachties, fishermen, pleasure cruisers or merchant seamen:

I had returned to Oban for a week to sit my Royal Yachting Association (RYA) Coastal Skipper's practical sailing course, during which I spent five days living aboard a local instructor's yacht. We navigated the varied waterways around the extended Oban area in a variety of conditions and using a variety of techniques. I learned new skills, visited new nooks and crannies of the coast, found new confidence in aspects of sailing which had hitherto intimidated me and generally had a very challenging but supremely rewarding experience, albeit one which was – perhaps understandably – somewhat less poignant than my approach in Gambo.

On the final day we were tacking back and forward up the western shores of Lismore Island when we overheard a mayday call over the VHF. A small fishing vessel, likely a lobster boat, was apparently sinking in the vicinity of the Island of Seil. We heard the hurried and somewhat truncated mayday, heard the coastguard's reply, and then nothing further from the stricken vessel. It seemed that it had gone down. What followed was a very memorable show of salty solidarity as what seemed like every vessel in the area headed for the scene of the accident, under the guidance of Clyde Coastguard, to search for the sole crewman who was now evidently in the water. It was April, and the waters were cold, meaning that the casualty had a very limited window of survival.

Gradually more vessels appeared on scene, including yachts, other fishing vessels, the tour ship *Hebridean Princess* and, the Caledonian MacBrayne ferry Lord of the Isles.

Now, as all of this was unfolding, still on VHF channel 16 I might add, both Oban and Tobermory lifeboats were scrambled. Unfortunately even from Oban the travel time between initial alert and arrival at the island of Seil was upwards of thirty minutes. So, the coastguard needed to nominate a vessel at the scene who could co-ordinate the other attending craft and act as a go-between. The Lord of the Isles, being the largest vessel, had the strongest radio, the highest vantage point and the most numerous bridge crew. And so, unsurprisingly, we heard the coastguard asking Lord of the Isles to step up. And so they did, without hesitation, despite having a full complement of passengers on board who at that time were in transit back to Oban from the Island of Colonsay. This would obviously interrupt their travel schedule in no uncertain terms, and with passengers who in all likelihood had onward travel arrangements and commitments in Oban. However, the ferry continued in the role of on-scene co-ordinator for some time, even after the arrival of Oban Lifeboat.

We listened to events as they unfolded for well over an hour. Oily patches were spotted on the water and reported via Lord of the Isles back to Clyde Coastguard. Some floating fuel cans were discovered and their locations relayed, all of which is invaluable information for the coastguard whom, using local tidal flows and current wind activity can plot up the most likely spot at which a casualty vessel went down or that a person in the water might end up. This search continued until eventually, completely un-prompted, Clyde Coastguard radioed the Lord of the Isles and asked something to the effect of:

"Lord of the Isles, we extend our thanks to you for your services. However we recognise that you are currently on service and have a schedule to keep. How long are able to remain on scene?"

There was barely a moment passed before the reply came back:

"Clyde Coastguard, Lord of the Isles. For as long as you need us Sir."

I have rarely been so moved, or had my faith in community spirit so forcefully re-affirmed, than by that single statement.

Tragically the casualty wasn't found, at least not in time. However, often good things can come from tragic events, and I for one took something positive away from what happened that day.

Anyway, back on Gambo everyone was still safely on-board and fortunately we gave the Lord of the Isles no reason to interrupt her transit on this occasion. And so, with cups of tea held firmly in our hands, we continued on our way eastwards, knocking off the miles like pints on a Friday night:

Waypoint #5: *'N. Salen Bay': 56° 32.99' N, 005° 55.98' W.*

Check!

From here on, the waypoints seemed to fall like dominoes:

Waypoint #6: *'Ardtornish Bay Beacon': 56° 31.08' N, 005° 45.57' W.*

Check!

Then, as the light of an overcast dawn slowly began to paint the world around us in a subdued shade of silvery grey Duart Castle's familiar picture-postcard face appeared to our starboard side and we approached the eastern end of the Sound of Mull:

Waypoint #7: *'Duart Point': 56° 27.49' N, 005° 37.25' W.*

Check!

At this point we turned slightly to starboard, avoiding the shallows and tidal race south of Lismore island and clearing the marked rocks beyond:

"On to 'Sgeir Dhonn'.

Check!

And there it was, the final short stretch of water beyond which lay our final destination: Oban. Already we could make out individual houses and cars moving about the Hillside roads. The last gasp was at hand! And so, as the early morning broke and the day's fishing traffic spluttered past us, heading westward and out into the seas from whence we had just come, we motored wearily past our final waypoint. *'Lady Rock'*, and headed into Oban Bay.

I was utterly exhausted, but familiar enough with boat handling in in-shore waters to know that we were still far from safe. Not until the mooring lines are in place and the engine has finally been switched off do I ever really relax. So, I ushered Sam and Matt up onto the bow to look for obstacles in the water ahead. No sooner had I done so than a brightly coloured lobster pot marker buoy slipped past us to port, vindicating me in my caution. Honestly, that would just be the icing on the cake! Another fouled prop within virtual shouting distance of Oban Lifeboat station! Oh the shame! At least when we managed to virtually incapacitate ourselves in the mid-Atlantic there had been no-one to point and laugh. Here we had an audience, Oban Bay being somewhat amphitheatre-like in shape; the perfect geography for viewing any maritime calamities which might pan out below the town's elevated terraces.

No, failure simply wasn't an option. We were virtually there dammit. I may have been awake for upwards of twenty-four hours by this point but I wasn't going to fluff it now.

I dropped the engine revolutions right down and nosed around the northern end of Kerrera island and into the bay proper, only to see two Caledonian MacBrayne ferries jostling for their place at the new and rather plush looking Oban ferry terminal. I had a brief but memorable panic, I'll be honest, not quite knowing whether the ferries were coming or going. They were a heck of a lot bigger than we were! So, over-cautious as ever, I killed the revs and sat motionless in the bay for a moment until I was sure we weren't about to be cut off. As the ferries manoeuvred themselves around the terminal berths and one turned to steam out of the bay I pushed the throttle down again and gradually brought us about and headed into the marina. This is where a wholly different kind of challenge always begins.

As terrifying as crossing an open ocean is, oceans are pleasantly notable for being somewhat empty 999 days out of one thousand. You rarely see other ships, as we can testify, and generally speaking the only things likely to wrap themselves around your propeller start their day on your very own deck. This we can also confirm. So, imagine the shock of suddenly moving into an area festooned with other vessels, all moored to ropes which lurk unseen beneath the surface. Suddenly the potential for a good old fashioned foul-up is limitless. So, it was with EXTREME care that I manoeuvred us amongst the assemblage of other boats, looking for somewhere to put in.

Eventually I settled for the quietest corner of the pontoon I could see. It didn't quite look big enough for Gambo long-term, but at that particular point, in the here and now, I didn't care. It would do.

Matt and Sam got the mooring lines together, rigged them up on the port side ready, and stood-by as I brought us in very slowly and nose first.

Slowly.

Slowly.

Slowly.

One quick, hard burst of power astern and . . .

Thud.

Hmmmm. . . . Perhaps not quite *"slowly"* enough.

After we gently 'nudged' the pontoon both the lads jumped onto the marina decking, wrapped the lines around the cleats until we were sitting comfortably on our fenders and . . . and nothing. I switched the engine off:

01/10/10
0640 UTC. Sam:

ARRIVAL KERRERA MARINA, OBAN. 0632 Engine stop. Engine use 11 hrs 31 mins. Trip total: 123 hrs 30 mins.

We made it. The grass IS greener on this side!

23 / Drying out

The first thing that happened after we arrived: Sam got land sick. It was incredible really, for he had been virtually bulletproof all the way over the ocean, but as soon as he stepped off Gambo, almost instantly in fact, Sam felt sick. Poor bugger.

The second thing that happened after we arrived: I took a group photo of us enjoying the grass. And I can tell you we REALLY enjoyed that grass. New smells! New colours! For a few moments we all just lay in it, spread-eagled on the first living vegetation we'd touched in eighteen days and almost 2000 nautical miles. I could barely even pull myself away to get my camera ready.

The third thing that happened: I went for a shower.

Later, after maybe a couple of hours of trying, unsuccessfully, to get some sleep, Sam, Matt and I hopped onto the little ferry which connected the marina to the mainland and wandered into town. Our primary focus was to announce our arrival to the authorities, but we also had some other burning matters to attend to. One of those issues was maybe better described not as a burning matter, but more as a stinking and chafing matter. For you see we rather badly needed new clothes. The rags we had on at that point (albeit concealed beneath heavy foulie jackets, which it simply hadn't occurred to us might be somewhat conspicuous around town in Oban) had evolved into being less like clothes and more like living, breathing members of the crew. I'm pretty sure they had become so heavy with organic matter that they could very well have told their own stories in plain English had anyone thought to ask them. This was a particularly pressing issue for Sam as later that night he was due to be meeting his then girlfriend's parents for the first time. So, intent on making ourselves at least *look* and *smell* human again, we strolled (through the midst of growing land sickness, by which Sam was being increasingly badly affected) along Oban's main shopping street and staggered into the first charity shop we could find. Here, between us, we scored two pairs of trousers, a belt, several well broken-in shirts and a woolly jumper.

The new clothes were on us before they had even been paid for and we emerged from the changing rooms looking like new men. Instead of handing our purchases over to the till operator to be scanned and bagged we presented ourselves at the cashier's desk already in our new attire and with several handfuls of price tags. I think a few items which we had previously been wearing found their way swiftly to the nearest street bin.

Next on the agenda was the train station where Matt bought himself a ticket for the first train back to Sheffield. He was already a few days late for the start of his Master's degree course at Bangor University in Wales and so was understandably somewhat anxious to get home. So, after a somewhat dazed saunter through the supermarket and an equally confused visit to the office of the local paper (where we had been asked to come along and give an interview) it was soon time for the crew to go their separate ways. Matt threw his kit together and was gone. Not long after that Sam's girlfriend appeared and took him away too, leaving me alone on the boat with a shopping bag of goodies. Apparently, what I had craved more than anything after weeks at sea was a sirloin steak, sweet potatoes, mushrooms, fresh milk and an avocado. I was planning on having a good dinner. Well, after I'd had a good sleep that is. I could barely see straight by this point having been awake for almost forty-eight hours. So, I put my head down.

My head stayed down for a long time. I was sleeping the deep sleep of someone who hasn't had solid, uninterrupted rest for many, many days. This coma was interrupted only by a phone call from Alun. My mobile phone came to life next to my head. It was the first time it had done this (ring, that is) in over a month. I answered, although I think my body performed this operation more on autopilot than anything else. Alun, who was waiting for his third child to be born at that precise moment in time, somewhat agitatedly asked me where I was. This is a simple enough question, but I was totally confounded. I honestly had no idea where I was. Try as I might, I just couldn't summon an answer for him. I looked around and saw the usual damp cushions, greasy plates, scattered pasta shells and worn deck boards of Gambo's cabin. Things didn't seem to be moving as much as they usually did though, so with the phone still pressed to my head I waddled to the companion way and looked out of the hatch. I can't remember clearly enough to repeat what I said, but to the listener on the other end it must have been a bit surreal. I think it went something like:

Colin: *"Whoa, OK. There's trees here. Uuuuuhhhhhh.... I think we must be in Oban. I'm not sure though mate. I can't remember. Honestly, I'm not sure. Errrrrr...."*

Alun: *"Where are the other two, Colin? Is Sam there?"*

Colin: *"Errrrrr.... No-one else is here. I'm not sure where they are. Uuuuuhhhhhh, I'm pretty sure they were here earlier, but it's just me now. I don't know why. S**t, I honestly have no idea what's going on...."*

After placating Alun and at least temporarily satisfying him that I was simply tired to the point of madness, and not on mind-altering drugs, I went back to sleep for a few hours before eventually getting up and cooking myself the dinner that I'd been so looking forward to. The next day I set about 'putting Gambo to bed' for the winter, stowing numerous things which had hitherto been secured on the upper deck down below under lock and key. These included a large 100-metre roll of one-inch diameter polypropylene rope, the (very) heavy life raft and a forty horsepower Yamaha outboard engine. A fair bit of heavy lifting all in! Matt and Sam and I had already pitched in together to wash and stow all of the foulies, life jackets, dry suits and other general sailing kit that we had used during the voyage.

Gambo was to sit alongside at the marina for several weeks now before being moved and lifted out of the water. She would then spend the worst of the winter months propped up on the marina's hard standing where she would await our return the following spring. So, knowing she was going to be left to her own devices for a while, I shifted the mooring lines, doubled them up and tied cut lengths of garden hose around likely chafing points so as to make sure Gambo didn't make an unexpected bid for freedom in the absence of any crew. I was actually slightly shocked to find that in only twenty-four hours or so, two of the lines were already beginning to chafe through. These had been rubbing against the steel gunnel through which they passed from the deck cleat to the pontoon. It *had* been pretty windy, but I was still surprised. It really is phenomenal how the wind and other elements will perpetually assail a boat. The second you take anything for granted, over-strain something or fail to react when you should to changing conditions, things break.

Eventually I managed to 'let go' sufficiently to start getting my own gear together and ready for my re-immersion into normal society. The following afternoon my mother and stepfather arrived to pick me up and drive me back to their home near Edinburgh. Their arrival provided me with an opportunity for which I had waited a long time: the chance to show my mother the boat in which I had spent so many 'character building' weeks and months of my life over the years. I was actually pretty excited about this, as Gambo had come to mean rather a great deal to me since I first set foot on her in Uruguay back in 2004. Metaphors aside, I had spilled rather a lot of blood, sweat and tears on her decks, cumulatively sailing her the equivalent of one half of the circumference of the Earth. Understandably, given the worry I know I must have caused her, my mother was quite reserved when she stepped aboard. My main memory of this one-off meeting was her less than enthusiastic reaction to the smell of the main cabin and her outright refusal to drink anything from teacups covered in engine oil. Ah well.... It takes all sorts!

In short order Gambo was locked up – her keys stashed away on deck – and, harbouring some very mixed and contradictory feelings of joy and sadness, I walked away from that boat, leaving her all alone on the pontoon. It felt somehow wrong.

For all our inexperience, all of our occasional incompetence and despite all of our mistakes, Gambo had taken care of Matt, Sam and I, carrying us safely across that vast expanse of ocean, one which was now far more tangible to me than it had been only a few weeks previously.

My parents and I were soon riding back across to Oban on the Kererra ferry. We headed for the car and no sooner was its door closed behind me than I was settling into the back seat like a cat into a pile of wool. As we drove back to their home in Midlothian I sank into the kind of relaxed state I had only dreamed of for the last few weeks. It was absolute bliss. For almost twenty days I had been jointly responsible for navigating a sailing boat, virtually no component of which would independently have floated, through 2000 miles of belligerent salt water. Every metre of ground won was wrestled from the hands of the wind.

Wind has to be managed, and as we had learned the hard way, it will make you pay dearly if your attention wavers. By comparison, being ferried home in a metal chariot, travelling at fearsome speeds along tarmac strips beset on all sides by other steel boxes moving at equally terrifying speeds, seemed positively safe. Nothing could possibly touch me now!

I went to bed that night in a clean, soft and entirely salt-free bed, and couldn't sleep. I lay awake for quite a few hours I think, unable to settle. Surely I ought to be doing something, or at least *worrying* about doing something? No? Apparently not. I forced my mind to calm down and eventually sleep overtook me. When I next woke it was perhaps six in the morning and the sun was shining down on me from somewhere. I sat bolt upright and looked around in confusion. Where was I? What was happening? In something of an uneasy daze I looked up towards the light which was pouring into the room and, to my absolute horror, I saw trees and buildings! The only thought in my mind at this point was that there should be no trees, nor should there be houses. I was utterly and single-mindedly certain that I was still on Gambo, and the sudden appearance of these dry-land objects could mean only one thing: We had crashed! I panicked and reached for the latch of the window, wholeheartedly believing it to be the companionway. I tried to open it but it wouldn't budge. It was locked. The key lay next to it, but that didn't seem right. Gambo's hatch didn't lock from the inside! Something wasn't right. For several long moments I sat and stared out of the window as reality trickled back into focus. It was slow and uncomfortable, but I began to understand that I wasn't on Gambo any more. I was at home, in my mother's house. I was safe and sound and healthy, at least physically. That was the moment that I finally realised that we'd done it.

Epilogue

To this day I am glad that the window to that bedroom was locked. If it hadn't been, I honestly believe that I would have opened it and climbed out, intent on somehow trying to save an imaginary sinking ship. That bedroom is on the top floor of my mother's house, and with only concrete slabs below to catch me it might not have ended well.

But what does all that mean? What is my final point? Well, I guess it's this:

Adventures are their own reward in that they let you access parts of your heart and mind which sleep when you are safe and comfortable at home. This kind of thrill can of course come at a price though. Danger and challenge can make you feel so wholly alive and deserving of your place in the universe that everything else seems mundane by comparison. And herein lies the problem, for everyone needs a home. The harder you chase, the more you leave behind. The more 'exceptional' you try to make your own life, the less readily it fits into the world which is built by others while you are away. People have shapes, and the more complex you make your own, the less likely it is to fit alongside another. Yes, boredom is poisonous, but loneliness can be far more dreadful to bear than the worst of any Atlantic storm. For why do we build careers, relationships and support networks unless for the benefits and sense of meaning they bestow upon our lives?

But still some of us are drawn away from the warm glow of normality. Adventure and exploration bring meaning too, and often in a far more visceral manner. It can be easy to forget from whence your happiness flows when life's tides slop lazily about your feet, but when the adrenaline of an ocean storm crashes all around you, saturating your senses, there can be no question of what makes that moment special. And as for relationships, adventure need only be experienced alone if you choose this path for yourself. Living a life 'off the map' needn't bring loneliness, whatever some might say. For me, certainly in post adolescence, it has been all about the people.

When examining the point of motivations, Frank Wild, right-hand man to Sir Ernest Shackleton on his famously ill-fated 1914 – 1917 'Endurance' expedition, admitted to being called *"off the edge of the map"* by *"little voices"*, metaphorically I'm sure. In contrast, Sir Edmund Hillary spoke in his autobiography of a desire to test his own abilities, and push human capability to the limit. I can certainly empathise with these sentiments, but the buck doesn't stop there. Doubtless I've ranted enough about the merits of challenge and the alternate perspectives on life which adventure can furnish you with, but when it's all said and done, I doubt there would be many adventurers out there who would disagree with me when I say that the single most rewarding aspect of pushing yourself in extreme environments and in extreme places, as part of a team, is the quality of the relationships you form along the way. Matt Burdekin, Sam Doyle and myself crossed the Atlantic together. We are three very different people, but we each have our own strengths. We worked together and found a way to mesh those strengths in a way that compensated for our collective weaknesses. There was no skipper. There was only a crew.

If relationships are only as strong as the foundations on which they are built, then I look forward to a lifetime of understanding shared with the two men whom I truly met on the vast, unfathomable and ultimately unbeatable North Atlantic Ocean.

We had tested ourselves, and passed, together.

The Crew

SAM DOYLE still lives in Aberystwyth. He finished his PhD a little behind schedule on account of the fact that, being a bit bright, he landed a staff job early, working as a full-time 'post-doctoral' researcher – a position usually reserved exclusively for those with a PhD already behind them. In this role he returns regularly to Greenland where he continues to pursue research into the ice sheet's sensitivity to climate change. He'd probably describe this work as 'fairly interesting'. These days however, he flies home from the Arctic, at least so far....

MATT BURDEKIN got home from the Atlantic in time to start his Master's degree in marine biology just a little later than his course mates. He went on to graduate on schedule and soon found work 'looking for feesh' with the Inshore Fisheries Conservation Authority in the north-west of England. Now Matt lives in Chamonix, France, where he climbs, explores, and recently became the first ever male member of the Chamonix Knitting Club.

COLIN SOUNESS, a glaciologist and seafarer with an interest in wild spaces, now works seasonally as a Polar Regions guide aboard ice-strengthened ships throughout the Arctic and in the Antarctic. When not at sea he writes about his experiences and pursues an ongoing research project investigating the record of ancient glaciations in the Kalahari Desert, South Africa (!).

BONUS: Short stories & more from the voyage

"Hitchhiking with the rich and famous"

On one particular morning, perhaps five days after I'd arrived in Kangerlussuaq (Greenland's aviation hub) an unusual jet landed and parked up on the airfield.

At this point, having been in 'Kanger' for almost a week, I was definitely beginning to lose the edge of urgency that life back home in the hustle and bustle of the UK brings. In places where very little happens, one's expectations of life become somewhat sedated, so the arrival of a new aircraft was a real event. At any other international airport in the world this might be a commonplace occurrence. But, in Kanger, the appearance of an unfamiliar plane quickly becomes the talk of the town. It might help if I explain a bit more about the comings and goings of Kangerlussuaq's flat concrete centrepiece.

The usual suspects on the Kanger flight line include: the large red Airbus A340 which Air Greenland operates to and from Copenhagen; the myriad smaller red Dash and Twin Otter turbo props which serve other Greenlandic communities across the country; numerous Air Greenland helicopters (A-Star and Sikorsky airframes) and, back in 2010 and 2011, a very distinctive charter jet sporting a large graphic of a certain moon of Saturn which kept a constant trickle of oil and gas workers coming and going back and forth from Edinburgh in Scotland. Although these familiar faces provided Kangerlussuaq Air Traffic Control with their bread and butter there are always occasional oddities: drifters passing through our one-cart town. For example, The US Air Force occasionally land their ski-equipped Hercules C130s in Kanger before and after they make supply runs to the camps up on the ice sheet. From time to time Russian Air Force jets (for example the impressive Ilyushin II-76MD, which came in while I was in town) land to refuel during Arctic patrols or supply missions. Even some smaller, shorter-ranged private jets land to take on fuel during trans-Atlantic flights. I was told that earlier in the 2010 season U2 had briefly stopped off in Kanger as their tour jet refuelled between continents! Anyway, as I said, new aircraft constitute major events on the Kangerlussuaq conversational calendar!

On this particular day the unfamiliar aircraft in question was a Boeing 767. It was a fairly sinister-looking bit of kit, painted in a two-tone pale brown and with slightly fewer windows than is the norm. She also had no obvious markings affiliating her with any particular country or company, but she did sport a menacing black stripe painted across the windows of the cockpit which looked for all intents and purposes like a highwayman's mask. Fairly intimidating really!

After landing she taxied off the runway, rolled to a halt and was met by one of the mobile disembarkation gangways. All pretty run-of-the-mill really, but when her door opened out walked not a stream of visitors, but a lone, well-dressed businessman who stepped almost immediately from the gangway into a waiting car and left for the harbour (located several miles to the west of Kanger airport). Next, two aircrew disembarked wearing smart, black uniforms and accompanied by three or four very attractive, alluringly dressed stewardesses, all with dark hair and tanned complexions.

The little entourage of crew and service-staff headed straight for the local restaurant: a Thai-inclined greasy-spoon takeaway on Kanger's main strip which specialised in Musk Oxen burgers and greasy French fries. This troupe were pretty cagey, and we didn't get much information from them. It wasn't until later when we tapped into the fountain of all knowledge, i.e. the very sociable Danish couple who worked in the flight control office, that we found out exactly who the businessy-looking chap that had disembarked first actually was. It was Roman Abramovich, the Russian oil tycoon and owner of Chelsea Football club. Of course it was!

The story was that when the Petermann Iceberg had calved off in the north of Greenland, Roman Abramovich had struck up a bit of a wager with his pal and fellow millionaire Bill Gates. The meat in this particular bet butty concerned which of the two, using a combination of their respective private jets and yachts, could reach the iceberg and have a look first. On leaving the airport Abramovich had transferred straight to his super yacht (the name of which we learned from the Harbour Master was *Pelorus*) and begun motoring westwards out of the fjord. A nice way to spend a spare long weekend!

So, there I was, sitting around that often-frequented coffee table in Kangerlussuaq flight ops with Martin, whom I had been intermittently killing time with over the last few days. Both of us were becoming bored and were keen for a bit of an adventure. Naturally, conversation turned to Abramovich and his amassed toys, one of which was cluttering up the airfield and another of which was cutting a hasty course through the waters of Søndre Stromfjord. Now, from having observed Pelorus sitting at anchor in the fjord, we knew that she had a helipad. From our observations we also knew that this helipad was empty. Take into account the particulars of our own heli-centred dilemmas, which included a permit deficiency for *flying* north, and you might begin to speculate as to which way our collective brain cogs began to grind as we supped our cuppas amidst the flashing lights and beeping sounds of Greenland's aviation nerve centre. There was no piece of paperwork anywhere which said we couldn't *sail* a helicopter north! If we could only get ourselves and the airworthy heli onto Pelorus….

The whole story was already writing itself in my imagination at this point. A trip into the high Arctic with Roman Abramovich on his personal yacht! Oh the possibilities! The antics! The hijinks that would inevitably ensue! In the strange emptiness of the Arctic, breathing the madness-tinged air of Kangerlussuaq, it all made a crazy kind of sense, in a somewhat

David Lynch-esque kind of way. The idea was so ridiculous that it seemed almost certain to succeed!

Gripped by a mixture of excitement and panic we abandoned our cups of tea.

Cue a quick visit to one of Martin's local chums to borrow a hand-held VHF radio, and after a few mental calculations on Pelorus' likely cruising speed, time elapsed since her departure, the range of a hand-held VHF transmitter and the lay of the local topography, we were speeding westwards to the end of the road in Martin's green Jeep Cherokee. There, on top of a hill, about ten miles west of Kangerlussuaq, lay a meteorological station and radar array. This was our best chance at getting a signal to Abramovich's yacht.

"Yacht Pelorus. Yacht Pelorus. This is Arctic Science Support, Arctic Science Support on channel one six, over."

Silence.

"Yacht Pelorus. Yacht Pelorus. This is Arctic Science Support, Arctic Science Support on channel one six, over."

Silence.

"Yacht Pelorus. Yacht Pelorus. This is Arctic Science Support, Arctic Science Support. Nothing heard on channel one six. I say again: Nothing heard on channel one six, over."

And so on and so forth…. To no avail.

It's more than likely that Pelorus, whose top cruising speed Wikipedia tells me is 19 knots, was already well out of range of our modest hand-held radio. Damn it!

We were gutted. My fantasies of what would undoubtedly have been the most eccentric hitch hike of my entire life slowly faded into the realms of the 'what might have been' and we were left with little more than a solid feeling of 'Well, at least we tried!'

Seriously though, how awesome would it have been if that plan had worked!? Incredible. We had been a touch over-optimistic I'm sure, but if there's anywhere on Earth that plans like that stand even the slightest chance of coming off, it's the Polar Regions. The Arctic, and the Antarctic for that matter, are both lands where the surreal is commonplace. It would have been amazing. Thank you Kangerlussuaq, for furnishing us with the chance to even try! I feel privileged to count myself amongst the select few who have even had the opportunity to hail Roman Abramovich's luxury yacht in search of a free ride! Ha. Next time!

"ALWAYS carry a spare tyre"

In stories from the American old west, be they word-of-mouth tales or big-budget screen shows, the drama often begins with an unexpected horse carrying an unexpected rider into a small, simple, near-horizon kind of town. Shots of an eerie desert, regurgitating a dusty, dehydrated mystery man into a surprised line-up of village folks spring to mind. The almost geological silence of the desert broken by approaching hooves, or heads turning along a parched yet squalid frontier high street as a stranger rides in. These are the kinds of scenes I imagine. By way of comparison, in Kangerlussuaq stories usually begin with the words "One day, a plane came in…" Well, after I had been in Kanger for perhaps eight days or so a plane came in from the town of Ilulissat (a couple of hundred miles north of Kangerlussuaq). This plane was carrying another guy from Aberystwyth University. His name was Mark Neal, and he was from the computer science department where he specialised in robotics. Mark had been working on Gambo with a radio controlled, floating sensor platform nicknamed "Minty", which the team were using to get sensors close in against the active calving fronts of numerous glaciers in the area. However, with both family and professional responsibilities back home Mark couldn't spend the whole summer in

Greenland (as much as perhaps he would have liked to) and so was heading home, on schedule. However, with the timing of flights etc., he would have to wait until later the following day to fly back to Copenhagen. So, not one to waste time and resources, the project boss enlisted him to conduct some "quick and straight-forward" laser scanning up at the edge of the ice sheet. I was to take some time out from my circumstantial incarceration and go along as an assistant. Happy days!

For our mini survey expedition we would need several things: a large and very expensive terrestrial laser scanner (TLS) which would build a highly accurate 3D map of any given site, a number of large batteries off which to run the scanner, a laptop and, by no means of least importance: a 4x4 vehicle.

As getting around Kangerlussuaq is pretty time consuming on foot (buildings, although few in number, are widely spaced), and carrying the TLS is pretty full on (those things are almost as weighty as they are pricey) we opted to get the motor sorted out first. We were in somewhat of a rush it should be noted, as it was now 9 am. Mark's flight to Denmark was at 6 pm, the scan site was a long off-road drive away and the clock was ticking. So, we headed to Kangerlussuaq International Science Support (KISS for short) to pick up a truck which had already been booked for us. KISS is a shoebox-shaped building on Kanger's main drag which acts as a bit of a hostel cum rental facility cum ad-hoc lab facility for visiting scientists passing through or working out of Kangerlussuaq. It's pretty friendly, and can be great fun towards the end of the season when researchers come back from the field, replete with new data, excited to go home and with the remains of the field camp's booze ration to be polished off before getting on the plane.

We arrived at KISS, signed the forms, picked up the keys and stepped outside to admire the battered, somewhat 'seasoned' Toyota Hilux pick-up which would be our chariot for the day. As designated driver I was over the moon! In years past I had owned various vehicles including, at separate times, two long wheel base Land Rover Defenders. The first one was a classic farmer's wagon which I had bought for a mere one hundred pounds (it'd had a cracked cylinder head which I'd known all about in advance) and the second was an ex British Military workhorse which I had run as my daily runabout for over a year before my return to university (and the consequent impact that life decision had made on my financial situation) forced me to sell it. I love 4x4s and off-road driving, but had never managed to get myself behind the wheel of the famously robust 1990s Hilux, until now that is. I just hoped I'd get the chance to properly test her out!

I really needn't have worried.

Anyway, so, now happily ensconced in our rusted, two-tone tank of a transport we headed off to do what needed to be done before departure into the field. We picked up the TLS and two twelve volt batteries from the University's lock-up, which was just down the road from KISS, and then headed off to fuel up, equipped with two extra Gerry cans which would give us as much running time out in the wilderness as we could possibly need. It was during this visit to the town's fuel pump, which actually proved to be very time consuming on account of the fact that you need what essentially equates to a Kangerlussuaq town member's card in order to get fuel, that we realised we were lacking a spare tyre. This was bad, and time was running out. After eventually achieving a full tank of fuel (which involved calling in one of the helicopter pilots to use his 'club card') we headed back to KISS in search of a tyre, only to find the place abandoned. There was no-one around, and we had no means of contacting the KISS manager out-with his office hours. Time slowly ticked by as we weighed up the wisdom of just going for it without a spare tyre. How often do you really end up needing one after all!?

With only seven-and-a-half hours left until Mark needed to fly out, and about thirty-five kilometres of off-roading between us and the furthest of our day's objectives, we decided to risk it and head out to the ice *sans* spare. Time just seemed too limited to waste any either waiting for or chasing the KISS manager around Kanger. Who dares wins! No?

So, off we trotted along the fairly abysmal dirt track which would lead us to our objective co-ordinates. We had three separate sites to visit and scan, all at points just short of the ice margin. We hit the most distant first, taking over an hour to reach it. Here, we set up the scanner and got her running. This in itself took the best part of an hour, and by the time the scan was complete the twelve-volt batteries I had picked up from the kit store were almost spent, which was unexpected. It turns out that cutting edge scanning equipment fairly drinks power! Fortunately, the Hilux was kitted out with two ignition batteries, just in case. So, we quickly removed one, wired it up to the scanner and set off again at full juice! This worked pretty well, as we could just hook the battery up to the truck again between sites and re-charge it as we drove. So, we worked our way from one site to the next, finally coming to our last co-ordinates which were situated on top of a smallish hill that overlooked the front of Russell Glacier. This spot was a couple of kilometres walk from the track, and up a fairly rough looking slope. By this point we were pretty sore from carrying the scanner and the batteries around the place on our backs, so we looked at the map and saw that there was an alternative approach from the other side of the hill which would take us only a few hundred metres from the site. The only 'X' factor was the quality of the road. The topographic map marked only the faintest of tracks. But hey, that's what Toyota Hilux's were built for, isn't it?

So, with a certain amount of trepidation, we set out on the fairly sizeable detour required to navigate to our alternative parking spot. Soon the track had disappeared altogether and we were driving on a sandy river delta. It was great fun, but I was becoming a little nervous about what lay ahead. We were getting farther and farther from anything resembling a road.

Eventually, vague tracks reappeared and we happily rolled up onto them. It became apparent pretty quickly, however, that these were not tracks made by your average 4x4. This was more Unimog terrain, or something substantially bigger than our little Hilux anyway! The going was exhausting, and very slow. I really fought hard to get the truck through without damaging her. At one point we came over a rise and found ourselves in a V-shaped sandy gully, perhaps seven or eight metres wide and with steep slopes on both sides leading down to a narrow vertex. The track ran on from the top at the opposite end of the gully. Not wanting to lose momentum in sandy terrain I tackled it straight off in the lowest range gear and at high revs, making the best stab I could. The whole truck slid sideways down the sandy banking and crashed into the vertex at the bottom. Everything lurched to one side, but I refused to let up on the gas and kept her moving. I was a bit shaken, but overwhelmingly relieved that the axle hadn't sheared. I revved her up and she soldiered on, recovering herself and pretty impressively climbing up and out of the sandpit onto the moderately better track overhead. We negotiated a number of sand traps similar to this one, and some gut churning, uneven rocky debris embedded in the tracks, but eventually made it to the ice front where we came across an almost perfect parking spot on short, even, grassy ground. I'm pretty sure this is where tourist groups are driven to in large Unimogs and from where they can view the inland ice. An Arctic off-road experience. It was all really a bit much for our rusty little Hilux in hindsight, but at that stage we were just pleased to have made it without any damage to speak of!

We parked up, unhooked the battery, slung the TLS and made for the hillock where we'd be scanning from. I somehow felt a little uneasy though, and felt compelled to check the tyres before leaving the truck. I did a circuit of the vehicle, kicking each wheel in turn.

"Perhaps it's my imagination," I thought to myself, "but that rear driver's side one looks a little flat."

Hmmmm. My heart beat a little faster, and my stomach tightened somewhat at the thought of a slow puncture paired with the ground which lay between us and Kangerlussuaq, but there was nothing I could do at that stage but hope for the best. So, choosing not to think about it too much just yet, I nipped off up the hill after Mark.

What followed was a blissful couple of hours spent basking in the beautiful but tenuous Arctic sun as the TLS did its thing. Sometimes in life we find ourselves 'at work' in amazing places, being paid to seemingly just enjoy ourselves. Not often perhaps, but it does happen. As a glaciologist who has worked in many remote and beautiful places I count myself ridiculously lucky to have many memories of feeling that way, and I'd encourage anyone who might be in a quandary about whether or not to engage with science to remember that even office or lab based research necessarily very often begins and ends with getting out there and getting to grips with nature. Sitting in the glow of September's last full-bodied rays, reflected so fiercely off the edge of Greenland's colossal inland ice.... This was one of the good times. The eastern horizon was a formidable line of unbroken white, the western horizon a more familiar blend of sky, mountains and water. We sat almost exactly on the boundary between the two; a nexus between two worlds, where the land emerged from the past, both metaphorically and, in a way, literally. The breeze moved the flimsy Arctic vegetation, the ice glistened, the melt water gurgled around the glacier's margins, the laser scanner whirred … everything seemed to be right with the world. We even had a friendly interlude when an alarmingly confident Arctic fox managed to sneak up behind us and get its head inside my backpack, no doubt drawn by the smell of the spicy sausage I had inside it. In the event he lost his appetite, although I still couldn't say with certainty whether it was the chillies in the sausage that put him off, or simply the effect of having had his head squarely inside my bag, which I'd owned for five years and had washed precisely never in all the time I'd been using it.

All was well. Snow buntings twittered away as we monitored the laser scanner and enjoyed the kind of clear, crisp, refreshing summer weather that you only find in the high latitudes. In environments like that worldly stresses just seem to seep out through your skin and vanish into the cool air, almost as air seeps from a recently compromised tyre, for example. Ahhhh... Yes. The tyre....

With the scanning complete we packed up and began picking our way back down the hill to the truck. Even from a distance I could see that we had a problem. My heart sank as I noticed the pronounced rearward tilt of the Hilux. My initial fears on leaving the truck parked had been well founded. The tyre was half deflated, bulging outwards on the ground-ward side like an unusually shaped butternut squash. This was bad. Worse than that, this was downright embarrassing! Two grown men, several seasons worth of fieldwork experience, in the wilds of the Arctic with a flat tyre, no spare, no pump, many miles from Kangerlussuaq, no VHF radio, no satellite phone and no mobile reception. Oh the shame! What was it I said before...? Who dares wins? Hmmmm.... Yes, sometimes perhaps, but I'm sure even the SAS (the British Army's Special Air Service), whose cap badge bears that very slogan, will concede that very often 'daring' can have somewhat less desirable consequences. Consequences like, for example, driving back into Kangerlussuaq two hours after finishing our scanning on three good wheels and one steel rim wrapped loosely in shredded rubber. As we clawed our way back into the town we left a single ploughed furrow in the dirt road behind us, just so as to leave no room for doubt on which truck it was that had driven out into the tundra without a spare. Just follow the trench and you'll find yourself by the shifty red Hilux outside KISS.

Never, EVER let anyone labour under the misconception that a PhD level education equates to a badge of intelligence. Education does not engender common sense. Quite the opposite in some cases! However, failings in common sense can, from time to time, furnish you with some great stories, and no-one likes any kind of story more than those which revolve around acts of colossal daring/spontaneity/stupidity. Tales of this kind are common in academic circles. Tales like the one contained within this book in fact!

Acknowledgements

I would like to take a moment and directly thank a handful of people and organizations to whom I was unable to do justice in the text of this book.

Firstly, Professor Bryn Hubbard of Aberystwyth University. Bryn was my supervisor and academic mentor, and were it not for his open-minded attitude and apparent faith in my academic commitment I would never have been able to take the kind of 'holidays' needed to undertake adventures like the one related in these pages. Thanks Bryn, for also being an adventurer in your own right and trusting me to get the job done when I came home.

Secondly, Maxime Denuville. It was Max who had been Gambo's first mate and who very sadly broke his hand just prior to our scheduled departure from Greenland. Along with Nolwenn, I had sailed with Max in Nova Scotia, Canada, earlier in 2010 and immensely enjoyed both his company and tuition. Thanks Max, for unwittingly preparing me for the Atlantic epic. I was gutted to hear you wouldn't be joining us.

Thirdly, and on behalf of Sam Doyle, Matt Burdekin, Miles Hill and George Ullrich, I would like to thank the Gino Watkins Foundation for the funding they granted Sam and his team, enabling them to undertake their climbing expedition on Uummanaaq Mountain. From a personal perspective, if this funding had not been forthcoming it is likely that the climbing expedition would never have been possible, and then Matt would never have been in Greenland in the first place, let alone available to help crew Gambo back to the UK. So, thank you Gino Watkins, for giving us Matt!

Fourthly, I would like to take a step back in time to the year 2003 and thank my undergraduate dissertation supervisor, Professor David Sugden. David was one of the people primarily responsible for getting me into glaciology in the first place and advised me brilliantly after I announced to him that I intended to travel, alone, to Greenland for the first time. Thanks David, for fuelling my academic imagination.

Fifthly, perhaps it goes without saying that we all – Matt, Sam and myself – owe a debt of gratitude to Alun Hubbard. It takes a special kind of genius to create the kind of opportunities that we have all benefitted from. Thank you Alun. You changed my life.

Finally, to everyone else amongst the ranks of our friends and family, a massive thank you. Thank you for all your support, patience and understanding. Coming home is only a joy for having all of you to come home to.

List of nautical and specialized terms, and abbreviations

1200Z: This is a notation of time. The numbers refer to the 24-hour clock, and the 'Z' denotes the universal time zone 'Zulu' (see below).

Alongside: The term 'alongside' describes when a vessel, of any kind or size, is tied-up and secured beside either a pontoon, marina, harbour, quay or indeed another vessel which is itself also secured to a harbour.

BATT: Battery.

Bearing: The direction in which the vessel is pointing.

Bow: The front, or 'business end' of any vessel.

Companionway: The hatch, or 'doorway' between the rear deck and the inner compartment of any small vessel.

Course: The direction in which the vessel is actually moving, taking into account currents and wind.

Cleat: A cleat is a mounting either on the deck of a vessel or on the quayside onto which ropes can be secured temporarily. Typical cleats have two prongs which protrude in opposite directions, standing proud of the deck, allowing ropes to be looped or twisted around them.

Clew: The clew of a sail is the built-in hole at the aftward end of the canvas through which a rope is threaded and tied, securing the sail to the boom and/or the sheeting line.

Dodger: The structure over the companionway which crew on deck can use to shelter from, or 'dodge', bad weather and waves.

DTD: Distance to destination.

E: Abbreviation of 'East'.

Foulies: Foul-weather clothing. Typically very thick, insulated and waterproof.

Furl: To furl a sail, or a rope, is to stow it away, generally inferring a coiling action.

Genoa: The genoa is the large foresail on a yacht such as Gambo. In Gambo's case it was the largest sail on the boat. There are several other names for foresails (e.g. jib etc...) but each one refers to a different configuration of canvas. In Gambo's case however, we had a genoa.

GMT: Greenwich Mean Time. This is the mean solar time in Greenwich, London, during the winter. This time reference is generally used by British affiliated bodies such as the Met Office, the BBC World service, the Royal Navy etc.... Note that in summer, the term BST (British Summer Time) is used in the United Kingdom, equivalent to GMT+1 hour.

GPS: Global Positioning System. This is a global, satellite-based network used for navigation and many other things, but generally, and in this book 'GPS' refers to our own on-board GPS navigation terminal.

GRIB: Abbreviation of 'GRIdded Binary', referring to a format of digital data record.

Halyard: The rope which runs from the top of a sail to the top of the mast and then down to the deck. This rope is used by crew for raising or dropping sails.

Keel: The streamlined board or fin-like hull section which protrudes downwards from the bottom of any vessel. This increases stability and efficiency as a vessel moves through the water. Keels are usually very heavy, keeping a vessel balanced and upright.

Kts: Knots, or nautical miles travelled per hour. One nautical mile = 1.18 statute miles.

Lee board: A lee board is a board or plank fitted across the cabin-ward side of a crewman's berth, designed to prevent him from tipping out of bed with the rolling motion of the vessel.

Leeway: Movement of a vessel caused directly by effects of the wind on the frame of the craft, i.e. not by the wind's effect on that craft's sails. Thus, typically, 'leeway' refers to side-wards motion of a vessel which ought to be moving forwards, perhaps sailing on a 'broad reach', i.e. with the wind coming from either the port or the starboard side.

Mizzen: A secondary mast fitted to some yachts. The 'mizzen' is located aft of the main mast, and is typically shorter than the latter. This configuration, with a main mast and a smaller mizzen would make a yacht a 'ketch'.

N: Abbreviation of 'North'.

NM: Nautical Miles.

Octas: A measurement of cloud cover.

PADI: Professional Association of Diving Instructors.

Port: 'Port' refers either to a harbour or to the 'left-hand side' of any vessel, i.e. the left-hand side as you would know it if you were standing on deck and looking towards the front of the boat.

Reefing (sails): Reefing a sail reduces its overall area. This generally involves pulling a sail down partially and securing the excess canvas in some way. To subsequently increase the sail's area again you would 'un-reef', shaking out the canvas you had previously secured and raising the sail to compensate. Reefs are numbered according to how much canvas is being deployed. The 1st reef refers to the initial stage of reefing, moving from full-sail into more conservative sailing. The 2nd and 3rd reefs describe increasingly more sail area being removed and secured.

Rigging: The various ropes and cables which keep any sailboat's mast and sails up and intact and allow the vessel's crew to control the latter.

S: Abbreviation of 'South'.

Sheet: A sheet line is a rope used to control the angle that a sail is presented to the wind. By 'sheeting in' one would pull the sheet rope, thus bringing a sail towards the centre of the vessel. 'Sheeting out' would pay the rope, and thus the sail, out, moving it's canvas face more perpendicular to the orientation of the vessel.

Shroud: A 'shroud' on a yacht or other sailboat refers to a very tight rope

(or cable) which runs from the mast down either the port or starboard side of the vessel onto the hull or deck. This rope is very tight and keeps the mast up by redistributing the forces generated by wind on the mast and sail.
Sloop: 'Sloop' describes a yacht with only one mast, but with both forward and aft rigging. A sloop also typically has only one foresail, otherwise it would be a 'cutter'. Gambo could be sailed rigged as either a sloop or a cutter, depending on whether both the genoa and the staysail were deployed, or only one of the two.
SOG: Speed over ground.
Springs: A mooring line specifically placed in order to prevent forward or rearward motion of a vessel alongside.
Starboard: 'Starboard' refers to the right-hand side of any vessel, i.e. the right-hand side as you would know it if you were standing on deck and looking towards the front of the vessel.
Stay: A very tight rope (or cable) similar to a shroud but running either from the top of or part-way up the mast to either the fore or the aft of the boat, the tension of which keeps the mast up by redistributing the forces generated by the wind and preventing the mast from snapping, most of the time.
Staysail: A triangular sail hoisted in front of the main mast and run up a secondary forestay, inboard of the foresail (i.e. genoa).
Stern: The back, or rearward end of any vessel.
Track: The word 'track' generally refers to the trail left behind a vessel as it moves forwards, through the water.
Transom: The actual sternward face of any vessel, though usually a small vessel such as a zodiac or the likes. A transom is flat, not pointed, and in the case of an inflatable boat would be the hard board upon which an outboard engine might be mounted.
UTC: Coordinated Universal Time. UTC is distinct from GMT in that it is based on an atomic clock and is thus free from any variation caused by the position of the sun. It is also less ambiguous than GMT, which is complicated by the factor of British daylight saving time during the summer and winter seasons (e.g. in summer the United Kingdom uses GMT+1 hour).

VHF: Very High Frequency radio, generally used for nautical purposes over comparatively short distances.

VMG: Velocity 'made good'.

W: Abbreviation of 'West'.

Waypoint: A single point on a navigational chart or map, located for navigational purposes. Generally, 'waypoints' refer to intended destinations along a planned route.

WP: Abbreviation of 'waypoint'.

Zulu: Basically the same as UTC.

ABOUT THE AUTHOR

Colin Souness is a glaciologist and seafarer with a passion for wild spaces. He grew up first in Scotland's west highlands and later in her rugged borderlands and has been inspired by wilderness areas from as early as he can remember; so naturally, Colin went on to study as a geomorphologist. Later, after first crewing a five-month sailing expedition to Antarctica and then joining the Royal Air Force for a while, Colin went back to uni for a PhD studying ice on Mars. Now Colin works as a lecturer and Polar Regions guide aboard ice-strengthened ships throughout the Arctic and in the Antarctic. He is a graduate of Aberystwyth University and The University of Edinburgh.

CPSIA information can be obtained
at www.ICGtesting.com
Printed in the USA
BVOW03*1203081117
499867BV00006B/46/P